PEOPLE AND ENVIRONMENT

Encyclopaedia of Environmental Pollution and Awareness in 21st Century Series

PEOPLE AND ENVIRONMENT

Edited by

V.K. Prabhakar

ANMOL PUBLICATIONS PVT. LTD.

NEW DELHI - 110 002 (INDIA)

ANMOL PUBLICATIONS PVT. LTD.
4374/4B, Ansari Road, Daryaganj
New Delhi - 110 002

People and Environment

First Edition, 2001

ISBN 81-261-0921-1

PRINTED IN INDIA

Published by J.L. Kumar for Anmol Publications Pvt. Ltd., New Delhi - 110 002 and Printed at Mehra Offset Press, Delhi.

CONTENTS

PREFACE

Today, the cry of 'pollution' is heard from all the nooks and corners of the world, and pollution has become a major threat to the very existence of mankind on this earth. It is the major challenge of our times.

The pollution of various resources has gone to such an extent that we are unable to breathe fresh air and drink fresh water. On one hand the advancements of science and technology have added to the human comforts by giving us automobiles, electrical appliances, supersonic jets, space crafts, better medicines, better chemicals to control harmful insects and other pests, etc., but, on the other hand, they have given us a very serious problem to face that is pollution. Pollution can be defined as an undesirable change in the physical, chemical, or biological characteristics of the air, water, or land that can harmfully affect health, survival, or activities of humans or other living organisms.

Pollution does not have to cause physical harm, but it may merely interfere with human activities. Chemicals spewed into the air or water from an industrial plant may be harmful to humans and other organisms living near the plant. If expensive pollution controls are required, the plant may be forced to shut down.

India faces serious problems of environmental pollution. Water pollution from untreated community and industrial wastes flowing into our rivers and streams, increasing use of fertilizers and pesticides, dumping of organic wastes etc., are issues that need to be tackled on the priority. Air pollution, long regarded as the bane of industrialised nations looms large over our country.

Here, we are presenting this *Encyclopaedia of Environmental Pollution and Awareness in Twenty-first Century* in fifty volumes with a view to make people conscious about the seriousness of pollution in the present time and its remedies.

Editor

1 Introduction

The environmental destruction that has followed is making the lives of people humanly impossible, particularly of women who have to fetch daily household needs from the immediate environment. To survive, the poor often have no alternative but to turn upon the very environment with which they lived in full harmony in the immediate past. Millions of people today see no alternative for survival other than selling firewood from forests—a new profession that has emerged with amazing rapidity. A people based development process—especially a women- based process would not only demand the restoration and enhancement of nature but would also bring about family planning and stabilise population.

In 1967, William and Paul Paddock wrote a book called "Famine 1975—America's Decision: Who Will Survive?" They classified developing countries, based on their population growth rates and food production potential, into three categories: 'can't be saved', 'walking wounded' and 'should receive food'. India suffered the ignominy of being classified as 'can't be saved' with the comment that it "is the bellwether that shows the path which the others, like sheep going to slaughter, are following". India was a country, according to the authors, to which the US should not have bothered either to provide aid or food. Fortunately, India's food production managed to keep ahead of population growth, which helped it through some of the most critical drought years like 1979. But the apparent stagnation in agricultural production, though in 1983-84 a new record foodgrain production of over 150 million tonnes was recorded—once again created despondency. The 1981 census results turned out to be even more of a shocker than usual. The Planning Commission had just put the finishing touches to its Sixth Five-Year Plan when the Registrar General of Census announced that the country had some 13 million more people than the number estimated by the planners and that the birth rate was not 33 per 1,000 people but around 36. The Planning Commission had to hastily restructure the Sixth Plan to bring it into line with the new figures.

The key question for expert and lay person is: can India's lands support such a large and growing population? This is where the 'carrying capacity' comes in. It is the number of people or animals that an area of land can support on a sustainable basis. Unfortunately, not one expert in the country has attempted to quantify the carrying capacity of the area under a single development block leave alone the whole country. Indeed it is extremely complicated and uncertain business trying to set a figure to the number of beings a piece of land can support. Take the case of Fiji where, in 1973, UNESCO and the United Nations Fund for Population Activities (UNFPA) launched a pilot project to study the population-environment relationships in the eastern is lands. The number of ecological variables were found to be enormous, putting paid to all efforts to put a figure on the small islands' carrying capacity. Only specific inferences were possible. For instance, eastern Fiji is caught in the stranglehold of declining copra industry and is suffering from poverty and emigration. If the colluvial soils which have been planted with increasingly unproductive coconuts for copra export were converted to farming of taro, a root crop, it was estimated that up to five times more people could be supported with this economy than if they continued with copra. Any talk of carrying capacity must clearly be marked with great caution. In a modern interdependent world, where trade plays a major role in the population supporting capacity of a particular area can, in fact, increased enormously by importing a few critical items like plant nutrients from outside.

Densely populated Europe today is heavily supported by the production of a range of items outside European countries, especially in the developing countries. The UN has never dared to estimate how much land in the developing world is today producing food and other items like timber for the developed world and how the quantity of this land has been changing over time. If such a study were undertaken, it would clearly show that, despite decolonisation in large parts of the Third World over atleast three decades, many times more land is being used in developing countries to meet the needs of the developed countries as compared to 1950 and today this land use is a major cause of environmental destruction across the Third World. For instance, the first major attack on India's natural forests came with the demand to establish tea plantations to satisfy colonial Britain.

In late 1983, the Food and Agriculture Organization (FAO) released an extensive study on the subject, 'Potential Population Supporting Capacities of Lands in the Developing World', the result of almost a decade's work that began with preparation of a world soil map. The picture of India that emerges is both exhilarating and sobering. Simply put, India has enormous problems but also has an equally enormous natural resource base to solve them with. Assuming a low level of inputs: local low-yielding varieties, no fertiliser use or chemical pest, disease and weed control, long fallow periods, no long-term soil conservation measures, high rates of soil erosion, steadily declining soil productivity, manual labour with hand tools as the only power source, high labour intensity, subsistence production and fragmented landholdings.

The question has troubled sociologists for years: How has India's caste system, so elaboratean estimated 40,000 castes rigid, hereditarily determined, hierarchical and oppressive a social structure, lasted for so many centuries? Madhav Gadgil, an ecologist, and anthropologist Kailash Malhotra feel that the answer probably lies in the discipline that the caste system brought to the use of natural resources. It was a system which, on the one hand, forced its members to share natural resources and on the other, created the right social milieu in which sustainable patterns of resource use were encouraged to emerge. In other words, it was a social system which both forced and cajoled the social being right from birth to adopt sustainable cultural mores. Within the caste system, birth determined a human being's occupation. An "ecological space" and its natural resources could only be used by a definite occupational group.

This "resource partitioning" helped to reduce competition and, hence, conflict among human beings over scarce natural resources, and to create the right psychological environment: the allottees of an ecological space developed sustainable patterns because they had no worry that their resources would be snatched away from them and probably also because they knew that if they exhausted the resources in their own space, they would not be allowed to use any other. Malhotra and Gadgil are quick to point out that even though the caste system lasts till today, it has stopped "resource partitioning". The growing monetarisation of the Indian economy, first encouraged by the invading British then by independent Indian in the name of development, has destroyed the ancient sustainable and shared patterns of resource use evolved over centuries.

Social tensions are increasing as people fight over scarce natural resources, and because the caste system continues even after losing it raison d'etre, it has become a source of social power and oppression. These conclusions emerge from 12-year study of rural society in western Maharashtra. Here the Western Ghats rise abruptly, some 50 km inland from the Arabian Sea, to an altitude of 1,000 meters to 1,500 metres. Their 15 km to 20 km wide crestline is marked with heavy rainfall, more than 3,000 mm a year and it supports low but stable agricultural and pastoral productivity. The simple society here, mostly marked by single clan settlements, consists of few sedentary castes of cultivators, pastoralists and huntergatherers, which include Gavils who are mainly cultivators, and Kunbis who are buffalo herders and shifting cultivators. Further eastward, the Western Ghats merge with the Deccan Plateau through a series of broken hills, and the rainfall in this tract falls to about 1,500 mm to 1,000 mm near the edge of the plateau. This tract supports rich agriculture and neighbours a complex rural society with a number of artisan and service castes.

Still further east, the rainfall decreases further, to less than 600 mm in a semi-arid tract of 100 km width. Agriculture here is practised only in the river valleys. Pastoralists and huntergatherers use large tracts of uncultivated land. Because the terrain favours sedentary existence during the four monsoon months, and uncultivated lands are available for grazing, several pastoral as well as nonpastoral nomadic castes have established their base villages in this semi-arid tract. They migrate to the higher rainfull tracts, both to the east and west of the semi-arid tract, for the eight months of the year. Among those who have their base villages in this tract are Hatkars, Tirumal Nandiwallas, Fulmali Nandiwallas, Vaidus and Phasepardis.

The first example of "ecological niche diversification" is exhibited by the Kunbis and Gavli Dhangaras living on the crestline of the Western Ghats. The Kunbis cultivate paddy in the river valleys, practise shifting cultivation on the lower hill slopes and hunt extensively, they barter their cereals for butter produced by the Gavlis and other goods. The only cattle they keep are for draught purposes. The Gavlis on the other hand, live on the upper hill terraces. Although they practise some shifting cultivation, their main occupation is keeping buffaloes and cattle. They curdle the milk, consume the butter milk locally and barter the butter for cereals from the Kunbis. As they get their protein supply from butter milk, they do not need to hunt. Thus, the cultivation of valleys and lower hill slopes is restricted to the Kunbis and of hill terraces to the Gavlis. Maintenance of domesticated animals and exploitation of all fodder and grazing is restricted to the Gavlis and the hunting of wild animals to the Kunbis. Another instance of niche diversification is provided by the three nomadic hunting communities of the semi-arid tracts of western Maharashtra: Tirumal Nandiwallas, Phasepardis and Vaidus. The primary occupation of the Nandiwallas is entertainment and fortune telling, that of the Vaidus dispensation of herbal medicines, while the phasepardhis are specialist hunter-gatherers.

Unlike the settled castes the Tirumal Nandiwallas and Vaidus also hunt in addition to their primary occupation. But the three castes use distinctly different hunting techniques and specialise in different prey species. The Tirumal Nandiwallas use dogs to hunt wild pigs, large carnivorous mammals (leopard cat, hyena, fox and aquatic animals (fish, crab, turtle), Vaidus use baited traps for smaller carnivores (mongoose, civet, jackal and cat), monitor lizards and aquatic animals. The Phasepardh- is specialise in black buck, wild pig and birds such as doves, quails, partridges and peafowl. Wild meat means a lot to these castes. Gadgil and Malhotra studied the Tirumal Nandiwallas in greater detail and found that the median annual biomass of hunted animal is 2,180 kg per family or Tirumal Nandiwallas who are thorough in their use of the hunted animals. They even eat portions like the brain and the viscera, normally rejected by other people. Almost two-thirds of the live weight of the prey is consumed. In terms of per capita daily consumption the median works out to 57 gm of protein and 850 calories from hunted wild meat. For the Tirumal Nandiwallas, the hunted meat provides a more than adequate level of protein and over one-third of the calorific requirements of an average person. What is striking is that while the hunting techniques used by the three groups are significantly different, none of them are sophisticated that they cannot be adopted by another.

It is difficult to define quite where "environment" begins or ends for women in developing countries. As childbearers, family caretakers and consumers; as food-products, fuel and water gatherers and users; as field, forest, factory and office workers, women are primary managers, and often preservers of natural resources. As activists and leaders women are involved in campaigns to promote environmental awareness and protection. Women's work is generally undervalued. As a result, women constitute a disproportionate number of the poorest groups of people and are victims of hunger, illiteracy, poor health, scarce social and technical services, inadequate population policies, and other consequences of poverty.

In addition, womcn's participation and influence is inadequately represented in decision-making spheres concerning environment and development issues affecting the quality of their lives. Because of their daily tasks in the care of the family and community, women in developing countries effect and are affected by their environment. Poverty requires women to exploit natural resources in their daily tasks, rather than conserving them. Similarly, environmental degradation restricts the ability of women to overcome poverty. Given this relationship between poverty and resources degradation, actions to further the interests of both will be mutually reinforcing.

The following overview is intended to illustrate the nature of the relationship between women and environment, how development activities have compromised this relationship, and how activities designed to address both women and environment are mutually reinforcing. Women play multiple roles in the family, the community and the economy. Women may produce food and income. In many cultures women are the holders and teachers of family values and traditions. Because of their special responsibilities in the family and community, women have the potential to influence community attitudes and activities towards the environment.

There are natural linkages between the daily tasks involved in carrying out these roles and the natural and human resource environments. Women are the major producers of food for domestic consumption. They are the man drawers and carriers of water. They procure almost all the fuelwood for domestic use. In urban areas, shelter, sanitation,potable water,and a host of other social services are critical activities for women. Women and men differ in their perceptions of their environment and its appropriate or potential use. For example, women may perceive a local forest as a source of food, fuel and medicine for household use while men may perceive the value of the local forest as the sale of the cut trees in the market. Additionally, because women are often dependent on free goods such as water, fuelwood and fodder,they have a special interest in environmental protection and rehabilitation.

Women are often key agents in maintaining and enhancing the quality and stock of communal natural resources. The environmental knowledge acquired by many women is both useful and detailed. Close daily contact makes women, more than men, familiar with properties and uses of the natural environment, for example, disease and drought resistant crops and tree varieties, efficient fuelwood, and medicinal plants.

A survey in Sierra Leone revealed that women could identify thirty-one (31) different forest products while men were only able to identify eight (8)[3]. Women are often very aware of the ecological relationships that affect their everyday lives, for instance, trees and soil stability, trees and water regulation, plants and microclimates,, etc. Women thus represent a valuable information source on the local environment. Women are not only resource managers. They are also the victims of environmental mismanagement and contribute to environmental degradation. Where the resource base is declining or degraded, it is women who must search further and longer for fuel and water.

It is women who are most easily marginalized onto unproductive lands. Increased time and labour burdens are often manifested by stress, ill-health and malnutrition. The consequent entrenched impoverishment in turn forces women to further exploit the resource base, often violating their own knowledge and the well-being of the community. Women's local environmental interests have been affected by inappropriate development. For example, macro economic policies (increased cash cropping, timber production, higher food prices, reduced social spending) set up complex interactions which marginalize women by, for example, removing resources from their management and control and/or affecting women's ability to respond to a disturbed or changed resource base. Projects and programmes have also failed to address the specific needs of women. Although women's role in productive sectors such as agriculture, water, forestry, energy, and human settlements is well documented, their concerns, needs and special knowledge are often underestimated if not totally ignored. For example, as primary food producers or labourers in agriculture they are often denied land, remuneration, credit, training and other supports which would improve their health, income and status in an ecologically sustainable path.

Even environmental protection and repair projects with specific social objectives often fail to recognize the differing needs of women and men; consequently, projects are planned with unequal benefits accruing to both. While it is evident that women in developing countries are often the victims and agents of environmental degradation, it is important to balance this view with the recognition that women are central to resource management and are participating in, and leading, ecology movements in many countries. One example of women's grassroots environmental activity is the Chipko Movement in India. This movement began as a response to the deforestation of the natural forest with the reforestation by a single commercial species, eucalyptus. Indigenous forests provided food, fuel, fodder, household utensils, dyes, medicines, and income-generating products. Replacement by a single species had drastically affected rural women's ability to maintain a subsistence household. In order to protect the trees, women embraced the trees as a human shield against the tree cutters, have been challenged in the courts and have defied their own families. The movement has subsequently spread to other parts of India and to other issues. As well as women's grassroots activity, there are some examples of donor-sponsored resource projects which have been enhanced by women's participation in the planning, design and implementation. For example, the Andhra Pradesh Social Forestry Project in India indicates that initatives to address women's needs and aspirations have accrued direct benefits to women and their communities and enhanced overall project success.

The process of women's participation, both as agents and beneficiaries, brings many benefits. Such benefits include raising the status and self-esteem of women. Economic gains may accrue from new income earning opportunities such as small-scale local enterprises (soil, fuel and water conservation, seed selection, waste recycling, local exchanges of indigenous knowledge). Material benefits extend to the well-being of the family and community such as increased food security, child nutrition, health and education. Women's participation in resource decision making contributes directly to the protection and rehabilitation of the environment. Furthermore, increasing the capacity of women to overcome their own poverty has associated social benefits.

As women's income rises, there is a corresponding decrease in fertility rates. Rapid population growth-strongly correlated to conditions of poverty, compounds the relationship between poverty and environmental degradation. Therefore, improving the status of women is fundamental to bringing about changes in population growth and its associated effects on the environment.

Current wisdom is to see women not just as victims, but as major local assets to be incorporated in the interests of better environmental management. Supporting women's role as environmental decision-makers requires recognition and validation of women's role in the development process.

Probably no other group is more affected by environmental destruction than poor village women. Every dawn brings with it a long march in search of fuel, fodder and water. It does not matter if the women are old, young or pregnant; crucial household needs have to be met day after weary day. As ecological conditions worsen, the long march becomes even longer and more tiresome. Caught between poverty and environmental destruction, poor rural women in India could well be reaching the limits of physical endurance. Biomass plays a crucial role in meeting daily survival needs in the vast majority of the rural household in the country: food, fish, fuel (firewood, crop wastes and cowdung, organic manure, green mulch and forest litter), building materials (timber and thatch) and medicines (herbs), are all different forms of biomass. Water is another crucial element in human survival and its availability is definitely related to biomass in a country where rain falls only four months in the year.

When the biomass in the surrounding environment disappears, water sources like streams and ponds also tend to disappear soon after the monsoon. An important aspect of this biomass-based subsistence economy is that it is mostly non-monetised. Water, fuel, fodder building materials and even food to a certain extent are all gathered freely from the immediate environment. Production and processing of biomass agriculture, forestry and village crafts based on biomass as raw materials are also the biggest sources of employment.

Unfortunately industrialisation, urbanisation and the advance of the cash economy have greatly affected the country's biomass base. The destruction of biomass with deforestation and devegetation—and its transformation away from rural and household needs and towards urban and industrial needs (planting of teak or eucalyptus in place of sal or other natural vegetation, for instance) is having a major impact on the lives of all those who live within the non-monetised biomass-based subsistence economy.

Marginal cultures like that of tribals and nomads are the first to face total social destruction. But women, not only within these cultures but in all rural cultures, face the maximum threat. The special difficulties that women face-their extraordinary work burden or lack of access to health care, for instance do not arise out of ecological deterioration per se. They are located in the sexual division of labour marked by a double work burden (at home and outside) and by the specific nature of tasks they do, and in the unequal distribution of resources like food within the household, which stems from women inferior status in the household and from a lack of control over cash and productive resources like land. But given this situation, environmental destruction exacerbates women's already acute problems in a way very different from men.

With the growing awareness of energy and environment problems, several efforts have been made in recent years by the government to promote new technologies like biogas plants, fuelwood plantations, fuel-conserving smokeless chulhas, latrines and handpumps. These can reduce the rigours of rural living, improve health and meet fundamental household needs. Srilata Batliwala has calculated that if cooking fuel and water were readily available close to the household in

Pura villages and the efficiency of stoves was improved to reduce fuelwood consumption, a saving of nearly 500 calories per day per woman could be affected. Yet success stories about the spread of these technologies remain few and far between. A major reason for this failure is that these dissemination programmes seldom involve women either when the technologies are being developed or when they are being disseminated.

Any technology that supplies cooking energy should obviously be directed towards women. For example, a community biogas plant was set up in Fateh Singh Ka Purwa, a village in Etawah district of Uttar Pradesh, the first such plant to serve as an entire village. Every household, rich and poor, received gas from it. Technologically this demonstration plant was success but socially it was a failure. It ran well for a year but soon the community started losing interest. Most of its members stopped providing cowdung and the plant more or less came to a halt. Male community leaders pointed out they were not interested in energy for cooking. They would rather have energy for power irrigation pumps, chaff cutters and milling machines. A report by Nirlep Malhans and Jyoti Sanghera in 'Manushi' points out that women were extremely critical of the plant. It was decided, without consulting them, that gas supply would be limited to two hours in the morning, from 8 am to 10 am. By then, women were already in the fields. Clearly, the plant's organisers had completely ignored their work on the farm. The women also complained that the gas did not even provide 25 per cent of the day's cooking, and they now had to look for wood as a substitute for dung cakes. Further, the new technology also increased the women's dependence on men: they had to depend on men even for routine cleaning of the burners. After the Fateh Singh Ka Purwa plant, the government has established several dozen community biogas plants as demonstration projects in various parts of the country. But not one is under the control of women. Because of the failure of community biogas plants to involve the 'community,' there are now several proposals to use biogas energy to develop small/ cottage industries, in order to improve their economic viability. In an energy-starved environment, if questions of economic viability are raised, women's cooking energy needs will probably be relegated to second place.

Water carrying is another extremely strenuous activity undertaken by women and it consumes an enormous amount of their time. Available information indicates that women spend long hours and traverse long distances, especially in the hill areas and in the arid and semi-arid parts of the country. In the deltaic plain, the time taken may be less than an hour a day. But this varies: in villages of Karnataka, a study shows this can be from one to 1.4 hours per household a day, whereas in villages of eastern Uttar Pradesh it can be as much as one to 3.9 hours a day on an average. Contrary to the popular belief that women enjoy fetching water and washing clothes in a communal place because they can gossip, economist Bina Agarwal points out that the few studies, that have bothered to look into this aspect, note that, in fact, women usually prefer to wash clothes alone, and seek to avoid time when washing places or wells are overcrowded. Vina Mazumdar, Director of the Centre for Women's Development Studies in New Delhi, points out: "Irrigation gets high priority in most plans for rural development, but the supply of water for drinking and washing receives scant attention".

M.S. Swaminathan, former Member of the Planning Commission, in his J.P. Naik Memorial Lecture, admitted: "If men had to fetch drinking water, then 23,00,000 villages would not have remained without provision of drinking water after 30 years of planned development."

Very few attempts have been made even to combine irrigation and domestic water supply schemes. They are typically conceived of as separate and independent projects. Bina Àgarwal points out that irrigation schemes can in fact have an impact on drinking water supply and, hence, on women as primary carriers of water, even if they are conceived of as separate projects. Seewage from canals can raise the water table and unless this is accompanies by increased salinity, this would replenish drinking water wells and have a beneficial impact. But in several parts of north and central India, the rise in the water table after the construction of irrigation canals has also led to increased salinity in drinking water wells in the vicinity of canals. These wells have to be abandoned leading to severe drinking water problems.

Deep tubewells, on the other hand, can soak up drinking water table and increasing work for women who then seek alternative and more inaccessible sources. If the drinking water wells of untouchable households falls into disuse, they may be left with no other alternative but to depend on the mercy of the upper caste wells. They obviously cannot purchase their own tubewells.

Bina Agarwal gives the examples of Bulandshahr in Uttar Pradesh, where the drying up of village drinking water wells in untouchable communities, associated with the sinking of deep tubewells in the vicinity, has meant that the untouchable women now have to depend entirely on the goodwill of the higher caste to pour out for them water they need for their everyday use. Lack of clean drinking water most affects the health of children. But among adults, women are more exposed to hazards of polluted water than men. Firstly they are the primary carriers of water, and secondly they wash clothes and utensils.

Moreover, since childcare is primarily the women's responsibility, when children get infected women are more likely to catch the infection than men. Even if a woman did not fall sick herself, she would have to spend more time caring for the sick. Poorer households will be clearly affected the most, as they have less access to clean water sources and are less likely to have help in domestic chores.

How does the extraordinarily tough work burden of poor women affect their health? Firstly, studies show that they eat much less than they should, which leaves them extremely malnourished and susceptible to disease. Secondly, repeated child bearing—children are earning members of the family has a major effect on their health. Thirdly, the work burden is so high that most women do not seek medical care even when they are ill or pregnant. Batliwala has calculated the amount of energy spent by men and women in their daily chores in Pura village and compared this to their daily food intake. The amount of energy expended by women is about 2,505 calories daily and by men, 2,473 calories. But when food intake is compared; men get nearly 800 calories more than their expenditure while women get 100 calories less. This difference in energy intake between men and women has also been found in a study by Leela Gulati of the Centre of Development Studies Trivandrum, in a typical agricultural labour household in Kerala.

On employed days the women's caloric intake fell short of the recommendations made by the Indian Council of Medical Research by 20 per cent, while the man's fell short only by 11 per cent; on unemployed days the respective shortfalls were 50 per cent for the women and 26 per cent for the men. But Batliwala argues that this apparently small deficit of 100 calories a day amongst village women in Pura must be seen in the light of other realities of village life and of women in particular. Environmental sanitation is so bad and access to clean drinking water so limited, that the vast majority of Indian villagers have worm infestations. These parasites can 'steal' as much as one-fourth of the total food intake. During pregnancy and lactation, women need an extra 500 calories to 600 calories a day. At any point of time, one-third of the women in the country are in this condition, but most of them continue to work without any additional nutrition. The woman is the last to get food in an Indian household.

In this constant grind of survival, several studies have pointed out that children begin to acquire an extremely important economic role. They do many crucial tasks like caring for younger children, fetching fuel, fodder and water and grazing animals, so that the adults can undertake waged labour. In Pura, for instance, children provide 29 per cent of the total time spent on fuelwood gathering, 20 per cent on fetching water and 34 percent on livestock grazing. With time, the importance of children cannot but grow. As Amulya Reddy and Srillata Batliwala, who have studied energy consumption patterns is Pura in detail, point out: "Children have become the unwitting victims of the continuing energy hunger in Third World villages.... a family below the power the poverty line, which does not have the means either to hire labour or purchase labour-saving devices, is compelled to meet its energy needs only by producing several children." But this will not only have important implications for the education, health and nutrition of children but also for the country's massive family planning programmes and the health of women.

Male migration, a common phenomenon in large parts of rural India, is an important factor affecting the work burden of women. A number of factors have led increased migration in search of job opportunities: population growth, increased pressure on land, and a technological transformation in agriculture itself that has promoted landlessness. Introduction of irrigation and high yielding varieties in specific parts of the country have created regional imbalances in agricultural development. Migration of agricultural labour from Bihar and eastern Uttar Pradesh to Punjab now takes place on an extensive scale. Similarly, in fishing villages in Kanyakumari district men have begun to migrate more and more. Kanyakumari has become a labour catchment area for mechanised fishing in Kerala and eastern Tamil Nadu. Fishermen from Kanyakumari now spend anything between two to seven months of the year on mechanised boats in Cochin, Trivandrum and Tirunelveli districts, sometimes going as far as Orissa and Karnataka.

The destruction of forests and grazing lands has put extreme pressure on tribal and nomadic people, forcing them into landlessness and migrating in search of jobs. For instances, tribals are migrating in large numbers from Dhenkanal district of Orissa to cities like, Cuttack, Bhubaneshwar and places like the Assam tea gardens. Very little research has been done on the effects of this migration on the women left behind. Most studies of rural-urban migration have emphasised the positive economic benefits which accrue to the village because of remittances by the emigrants. But Andrea Menetee Singh of ILO points out that life of women left behind in the village can be hard. "If the oral tradition is any indication (folk songs, for instance), the emotional price the women pay for their husbands' emigration is very high," says Singh. Divorce rates are high among emigrants, which means that the proportion of women deserted after emigration may be fairly high. Surveys in Bombay slums have also shown that men who have migrated without their families have a high incidence of venereal disease, which probably means that they are transmitting these diseases to their wives on home visits.

Collecting fuel is an important daily chore for most rural women. In Pura, to gather 1.74 tonnes of firewood every year, each firewood gathering household spends an average of 2.51 hours daily, makes 172 trips in a year and each round trip is about 8.54 kilometres. In several north Indian households, there is greater reliance on cowdung than firewood. But this done not necessarily reduce the time spend by women. A study carried out on the health of pregnant women in two villages in western Uttar Pradesh show that women spend long hours in animal husbandry and making dung cakes.

For instance, one pregnant woman spent a 14-hour working day, in which three hours went on animal husbandry and half an hour on making cowdung cakes. In addition, this woman spent three hours on fetching water, 2.5 hours on cooking and five hours on other house hold chores, including milling grain.

Women's work in the hills and deserts is even harder. Agricultural production in these areas is extremely limited and alternatives to fuelwood like cowdung and crop wastes are also limited. Available studies clearly show that women put in an inhuman amount of time in the hills and deserts just searching for fuel and fodder.

2 Environmental Health

Environmental health is that aspect of public health that is concerned with those forms of life substances, forces and conditions in the surroundings of man that may exert an influence on man's health and well-being. This definition includes other people as part of man's surroundings that contribute to the status of environmental health.

Where not only do man and his environment interact, but man is shown to be a vital factor of his own environment. Early definitions of public health reflected concern with communicable disease and stressed prevention of disease. While prevention of undesirable conditions and consequences is fundamental to most definitions, public health, or just plain health, today has a much broader connotation including the attainment of desirable positive health values, in addition to the prevention of undesired negative ones.Consequently, the definition of public health developed by the World Health Organization, and widely accepted states. "Health is a state of complete physical, mental, and social well-being and not merely the absence of disease or infirmity."

Today, environmental engineering is a term that has come into wide use. While the meaning of this terminology is some what controversial is some circles, it generally refers to the study of man-environment interrelationships and the modification of the environment for man's benefit. To the extent that there is concern for man's environment as it relates from illness, health maintenance, human efficiency, comfort and the enjoyment of life, use of property, and management of resources. At times, it becomes impossible and useless, because of direct and indirect interrelationships, to make any distinction between those activities concerned with man's health and other aspects of environmental engineering and science, as in air and water pollution. Both natural and social sciences have application in the development of a suitable environment for man.

"Sanitation is a way of life' according to the National Sanitation Foundation, Frequently, throughout history, the status of the environment, which historically has been equated with sanitation but is now much broader in concept, has been the measure of civilization. In fact, the aggregation of population has always seemed to create environmental health problems which had to be solved before the populace could survive and enlarge. The earliest public health measures were concerned with environmental matters. Many customs, taboos, and religious practices has their roots in observed or presumed interaction between man and environmental conditions. Moses 'Law has dictates concerning sanitation which could have applications today.

The supply of water and the removal of wastes have been critical problems for aggregations of people. Drainage systems were built by the Minoans (3000 - 1500 B.C.), and the Egyptians (1000 B.C.) were noted for irrigation and other public health measure. The Grecians were concerned about personal hygiene and cleanliness. The Romans built aqueducts and drainage systems which were marvels. Archaeological excavations in India and the Middle East show that other ancient civilizations had similar systems, although not as impressive as those in

Rome. In the Middle Ages, there began a change in social organization which initiated the rise of many cities, with the growth of these cities, diseases were more easily communicated from one person to another. Terrible pandemics swept through the cities of Europe, decimating the population. Such diseases as cholera, associated with polluted water, and bulonic plague (the "black death"), associated with rats, were rampant. The very existence of the cities was threatened. With the Renaissance and the development of the miscroscope, it became possible to identify the bacteria that caused communicable diseases.

The first depiction of bacteria was by Anthony van Leeuwenhoek in 1683. Through the work of Pasteur, Koch, and others knowledge of the role of bacteria as a cause of disease was developed and enlarged. Interestingly, the earliest applications of Pasteur's work had commercial values in fermentation and animal disease. Before science established the cause of disease, man had, at times, instinctively avoided circumstances suspected of causing disease. Thus an epidemic of cholera and typhoid on London in 1854 was terminated by the simple expedient of removing the handle to a pump on a well located adjacent to a sewer.

The term "malaria" was the name applied to a disease thought to be associated with bad air, which was equated to the night air. We now know, of course, that the female anopheles mosquito attacks man and transmits the causative organism. Unfortunately, many of todays environmental hazards have been introduced so rapidly that man's instinctive reaction have not had time to adapt so he is not aware of their dangers. Nevertheless, such designations as "New Oriens" or "Yokohama" asthama, may well be the same intuitive labeling, witness "miners consumption" later found to be silicosis. With the coming of the Industrial Revolution new hazards to man were introduced into his environment-physical injury from machines and work methods, toxic exposures, and other stresses. It was about this time that the first organization of public resources and society in the interest of public health began. Developments in England and the United States were somewhat parallel.

Although they did not result in immediate achievements, in both cases social organization for public health can be traced to two reports. In England it was the "Report on an Inquiry into the Sanitary Conditions of the Labouring Population of Great Britain" by Edwin Chadwick in 1842. Chadwick's report revealed that one-half the children of the labouring population died before their fifth birthday. At that time he found that, in some cities, the average age at death was 36 for gentry, 22 for tradesmen and 16 for labours. Chadwick became interested in health as it related to poverty, and he was Secretary of the Poor Law Commissioners at the time of his investigations and report. In 1850 in the United States, the "Report of the Sanitary Commission of Massachusetts" was published. Lemuel Shatuck, a bookseller who became interested in statistics, was head of the commission. These two reports are hailed today as historical landmarks, not so much because they were forerunners of actions ultimately taken to establish organized public health agencies, but because of their insights and perspectives.

In Shatuck's report, we read the following:

> We recommend that provision be made for obtaining observations of the atmospheric phenomena, on a systematic and uniform plan, at different stations in the Commonwealth...
>
> We recommend that measures be taken to prevent as far as practicable, the smoke nuisance....

> (Noting that)...the tendency of our people seems to be toward social concentration...(this report continues):
>
> We recommend that in laying out new towns and villages, and in extending those already laid our, ample provision be made for a supply, in purity and abundance, of light, air, and water; for drainage and sewerage, for paving and for cleanliness.

Additional recommendations related to housing, schools, occupational health, and adulterated food and drugs. The report also urged..." that person be specially educated in sanitary science..." Now, over 100 years later, these environmental problems continue to plague us and in some instances, are intensified.

Life support	*Man's Activities*	*Residues and Wastes*
Air	Home	Solids
Water	Work	Liquids
Shelter	Transportation	Gases

Environmental Hazards

Biological	*Chemical*	*Physical*	*Psychological*	*Sociological*
Animal	Poisons and	Vibration	Stress	Overcrowding
Insect	Toxins	Radiation	Boredom	Isolation
Microbiolo-gical	Irritants	Forces &	Discomfort	Anomie
		Abrasion	Depression	
		Humidity		

Fig 1. The environmental system

The status of man's health represents the result of complex interactions between his internal biological system and the total external environmental system. This text is concerned primarily with the external environmental system, but it should be obvious that the interactions between the environment and man's biological system must always be before us. The environment is both natural and modified by man's work and presence. The element of the environmental system of concern are diagrammed in Fig. 1.

Man as a biological organism has developed in an environment which is being changed more rapidly than ever before. These changes may tax his adaptive capacities. Even where the adapts, we are not certain what the costs to his biological system may be. Certain environmental conditions are necessary to sustain life. Still other affect body functions, performance, and comfort, air, water, food and shelter are requirements for man's survival. Normal air in the troposphere contains 21% oxygen and 78% nitrogen. As the concentration of oxygen decreases below 16% anoxia develops which affects body functions and brain centres.

Life cannot be sustained when the oxygen concentration is below 6%, and immediate loss f consciousness results from exposure to a zero oxygen atmosphere. Air may also be the means for dispersal of toxic gases and particulates, microorganisms, and allergens. Air has been considered an unlimited resource, but experience has shown the natural dispersion and cleaning machinists can be overtaxed in extensive localized areas. As the population multiplied and man's activities pollute the air in increasing quantities, there is potential danger of overloading the earth's atmosphere. Survival in individual cases without water may be measured in terms of days (depending on other factors); however, civilization cannot develop or prosper without ample water for domestic, agriculture, and commercial activities. Many would specify recreational uses as a necessity. As with air, water may be the medium for transporting micro organisms and toxic chemicals.

In many areas the quantity of water available is a serious problem, but in others water resources have been degraded beyond use by pollution resulting from man's activities. The attainment of an ample and proper diet is a serious health problem in many places throughout the world. Even the United States with its abundant food surpluses has found that in segments of its population 20-60% of the families had diets that fell short of providing the minimum requirements for adequate nutrition state is free from hunger.

Expanding populations threaten to convert some of the best agricultural lands into sub divisions and industrial parks, which could have long-range implications for the food supply. Foods may be the mechanical means for conveying microorganisms and toxic substances.

Certain foods are also, under some circumstances, the media on which bacterial growth is sustained, with a resulting production of toxins. In many, if not most, areas of the world, shelter is necessary for man's survival. Shelter provides protection from excessive cold and heat, from the ravage of storms, and from environmental hazards. Properly designed shelter provides the spaces, services, safety, and privacy necessary for large numbers of people to live in close proximity to one another and in locations that would otherwise be uninhabitable.

As man engages in the activities associated with living, wastes are produces. these are products which have no apparent useful purpose, or they are of such marginal utility that recovery is uneconomical. Such products include human, residential, agricultural, commercial and industrial wastes of all kinds. the continuous removal and safe disposal of these wastes is essential to the continued existence of any community may be overwhelmed and its citizens endangered. These wastes may be solid, liquid or gaseous. Bodily discharges have historically been considered to be very hazardous to mankind. Intestinal diseases are readily transmitted where water or food is contaminated directed or indirectly by human wastes. Such wastes also provide a medium for fly breeding. Food scraps and waste food constitute garbage. When combined with other household waste, residential refuse is generated. The organic portion will attract and sustain flies and rats.

Where wastes are water-carried, pollution of water supplies may occur. Commercial and industrial liquid wastes may contain particulate and chemical pollutants. Many activities of man—driving an automobile, generating electric power, processing chemicals and petroleum, manufacturing certain products, disposal of wastes - may result, if uncontrolled, in the discharge of solid particles and gaseous pollutants to the atmosphere. Excessive pollution of the air has been cited as a cause of disease and death.

Aside from deficiencies in the life support system, man is subjected to a variety of environmental hazards. Some of these are natural, but increasingly environmental hazards result from man's activities and numbers. Sometimes man-made hazards are direct in their impact on other men; but they may be indirect in their influence, acting through other biological systems or overburdening the capacity of natural systems for renewal, dispersion, or assimilation.

The following factors can be used in categorizing environmental hazards: (1) biological; (2) chemical; (3) physical; (4) psychological; and (5) sociological.

Biological hazards are concerned primarily with the entry of disease producing infectious agents into man's body. Communicable diseases may be transmitted directly from one person to another by direct contact or aerosls discharged by sneezing or otherwise. Some times body discharges contaminate food or water consumed by the next victim or infect some other article with which the person has intimate contact. Where the environment may be modified to break the route of transmission from one person to another, such diseases may be prevented..

Animal may be susceptible or the carriers of diseases (zoonoses) which may be transmitted to other animals and some of these diseases may be transmitted to man in a manner similar to the transmission of communicable diseases from man to man.

Diseases are also transmitted from person to person or from animals to man by means of other biological agents. In such cases, these agents are disease vectors. Among such diseases vectors are flies, which may mechanical, when they puncture the skin for a blood meal and later infect the victim in a similar manner. In the case of the mosquito, however, part of the life cycle of the disease-producing organisms (as in malaria) may take place in the mosquito before another person may be infected. Other common disease vectors include fleas, ticks,mites, cockroaches, etc.

Man is susceptible to invasion by worms which may produce dehabilitation or produce symptoms of disease. Such entry occurs through the ingestion of the eggs of the worms, or through penetration of the skin by the worms. Certain diseases are characterized by skin lesions and pustules. Organisms contained in these lesions may be deposited in ruptures of the skin of other persons or on vulnerable mucous membranes by direct intimate contact. Examples of such infections are syphilis and staphylococcus infections.

There are other diseases which generally involve close person-to-person contact. Among these are infections of the upper respiratory tract, such as pneumoia, tuberculosis, and common colds, Aerosols composed of droplets discharged upon coughing, sneezing and breathing convey the organisms from one person to another. This transmission may be organisms from one person to another. This transmission may be aided by the use of common articles, such as drinking glasses, or by the contamination of foods. Ordinarily, bacteria require moisture, a food medium, and proper temperature and pH to survive any length of time. Drying tends to kill the bacteria.

Host Source	*Modes of Transmission*	*Means of Entry*	*Receptor*
Man	Direct contact	Skin lesions	Man
or	Aerosols	Nose or Mouth	
Animal	Waters	Food or Water	
Arthropods	Skin puncture		

Fig. 2 Modes of transmission of communicable diseases.

Communicable diseases are transmitted by the use of contaminated water for drinking, ablution and the preparation of food. Water supplies become contamined by receiving body discharges of man and animals. Bacteria are normally filtered out of water as it percolates through soils; however, water which follows passageways dissolved in limestone or fissures in rock formations will not be filtered. Formations composed entirely of coarse gravels will also pass bacteria.

Since many of the water-borne illnesses are of human rather than animal origin, the improper disposal of human wastes is the chief cause of contamination of the water. This may be from surface washings draining to a surface stream or entering underground supplies in limestone formations through sink holes. Shallow wells may become contaminated because the distance travelled through the soil is insufficient to filter out the bacteria. Wells which would otherwise produce safe water, by the use of contaminated water to prime pumps, or by contamination of ropes or buckets by unclean hands. Safe treated water may be contaminated through open reservoirs and by cross connections between the water pipes and waste piping. In the absence of organic matter to serve as food for the bacteria, there is a tendency for bacteria to die away in water. Water-borne illnesses include hyphoid and paratyphoid fevers, cholera, bacillary dysentery, amoebic dysentery, and infectious hepatitis. The microorganisms causing these diseases are found in the intestinal and urinary discharges of infected persons. Some times persons may be carriers of these symptoms of the disease.

Some of the communicable diseases generally transmitted by other means may be transmitted by water, e.g., undulant fever (brucellosis) and tularemia. Poliomyelitis is suspected of being capable of transmission by water. Swimming or wading in water may provide the contact for the entry of certain helminths, e.g., dracotiaosis (Guinea worm disease), echinococcosis, and scistosomiasis. The helminths are usually encountered in the trophis, but scistosomiasis has been a problem in some waters of the United States. Food may be the means for the spread of infectious diseases in several ways.

Consumption of the flesh of the infected animals, fowl, fish, or shellfish may result in the onset of a disease. Food may simply become contaminated by spraying with aerosols produced by coughing, by handling with hands or equipment contaminated with disease organisms, or by incorporating contaminated water with the food. In some cases, food acts as a medium for the growth of disease organisms, producing large numbers of bacteria or toxins.

Foodborne infections include such diseases as diphtheria and tuberculosis, as well as the intestinal diseases like typhoid fever and salmoculosis, dysentery undulant fever, Q fever, infectious hepatitis, leptospirosis, and botulism. A storage or holding period after contamination of the food during which the bacteria multiply is not required in these cases, although some of the foodborne infections mentioned result from illness in the animal, e.g., the contamination of milk as a result of bovine tuberculosis or undulant fever in cattle. Salmonella may increase in such numbers after contamination of the food that gastrointestinal symptoms appear following ingestion. This is a food infection.

Such foodborne infections are sometimes referred to as *food poisoning*, although this is not entirely correct. Strictly speaking, bacteria-caused food poisoning results from the growth of the bacteria in the food over a period of time during which the bacteria manufacture a toxin. When the food is ingested, food poisoning or intoxication results. Organisms frequently implicated are dstaphylococci and Clostridium perfringens (welchii). *Under anaerobic conditions,* Clostridiu, boutlinum *produces a fatal toxin which is not heat stable.*

Parasitic infections may be transmitted by foods which have contaminated. Diseased animals may be source of such infections if the meat is not thoroughly cooked, e.g., tapeworm and trichinosis. As may have been observed, some diseases may be transmitted through public water supplies, milk supplies, or other food.

Characteristically, water-borne epidemics affect families on the distribution system, or in the system contaminated. Similarly, milkborne disease out-breaks tend to follow the milk routes, with additional cases where purchases are made at grocery stores or restaurants. Outbreaks due to the consumption of other foods will be scattered depending upon the residence of patrons of restaurants, guests at private diners, attendees at picnics, and patrons of certain stores, according to the food distribution system and the point at which contamination occurs. Preliminary judgments may be made by mapping the residence of cases. Finding may be confirmed by taking case histories to identify common contacts and by bacteriological analyses of the water, milk and other food suspected.

In recent years, the filtration and chlorination of water supplies and the pasteurization of milk have virtually eliminated water and milk as means for transmission of disease. Where disease outbreaks occur from water or milk customarily treated or pasteurized, there is usually a human or mechanical failure resulting in inadequate or no treatment, or there is gross contamination following the treatment process. Such sporadic outbreaks indicate the need for constant vigilance for the safety of water and milk supplies. In contrast to recent experience with water and milk, food poisoning outbreaks continue at an undesirable level. There are a number of diseases of animals (zoonoses) which may be transmitted to man. This may be direct attack where the infectious agent is implanted in the wound. Rabies, rat-bite fever, and cat scratch fever fall into this category. Rabies, commonly associated with dogs, has been a hazards from bats and foxes and could involve other animals.

As indicated in the discussion of food, man contracts some animal, as with brucellosis, Q fever, bovine tuberculosis, various tapeworms (beef, pork, fish), and salmonellosis. Other animal diseases may be transmitted by direct contact, such as orthinosis (parrot fever or psittacosis), and by abrasions becoming infected, as anthrax (through skinning rabbits, etc.). Some diseases are conveyed from animals to man by anthropoids. These will be discussed under vector borne disease. It will suffice here to identify in this category the encephalitides, plague, murine typhus fever. Rocky Mountain spotted fever, relapsing fever, and tularemia. Obviously, some of the above diseases may be transmitted by more than one of the methods mentioned and sometimes by contact with waste discharges or food and water contaminated by such wastes. Vector-borne diseases constitute threats throughout the world. While any animals may be considered disease vectors, this discussion is aimed at diseases carried by arthropods.

With rapid transportation by airplane, arthropods may be carried everywhere unless strict precautions are taken. While much progress has been made in the United States in controlling such diseases as malaria, murine typhus fever, yellow fever, and bubonic plague, conditions exist which wound foster transmission if persons from other parts of the world, ill with any of these diseases, enter the community. Other vector-borne diseases do persis, such as the encephalitides, tularemia, and Rocky Mountain spoter fever. Plague exists in wild rodents in the sylvatic form in western states. Arthropods transmit diseases from animal to animal and from animal to man. By far the most important is the injection of the microorganisms in a wound made through the skin in the process of obtaining a blood meal. The mosquito is involved in this fashion in the transmission of malaria, yellow fever, dengue, filariasis, and he encephalitides, and similarly the rat flea in plague.

Other diseases are transmitted by fleas, mites, and tics where organisms in feces deposited on the skin after a blood meal or from crushed body parts are rubbed into or otherwise gain entrance to the wound. Epidemic typhus, murine typhus, scrub typhus, relapsing fever, Rocky Mountain spotted fever, and tularemia may be transmitted in this way. Arthropods may also be involved in the transmission of disease by simply transporting the organism from a source of infection (deposits of urine and feces) to hands, mouth, foods, or articles with which these come in contact. Such is the way that flied and roaches are involved in the spread of intestinal diseases-shigella dysentery, for example. Contaminated soils offer the possibility of infection. Walking baefoot or lying on contaminated ground may provide the opportunity for entry of hookworm through the skin. Wounds and abrasions may be contaminated in the way, as well as others, with tentanus.

Some persons are more seriously affected by communicable diseases than other. This may be related to their general state of health, concurrent diseases, physical vitality, natural or acquired immunity, possibly inherited factors, diet, environmental conditions, and other factors prevailing at the time of exposure. Thi is why some persons will become carriers, but not show symptoms of the disease; other will be ill to varying degrees; and still others die as a result. For many communicable diseases, immunizations have been developed which are highly effective in preventing illness completely or at least in minimizing the severity of the disease. However, besides being costly, it is difficult to maintain immunization in the total population.

Modern advances in medicine have resulted in a wide range of bactericidal therapeutic agents which are effective in abating disease in many instances. Unfortunately, some strains of the microorganisms develop resistance to these biological, rendering them ineffective, and in some instances persons develop allergies to certain biological when they have heavy or repeated doses. Thus the time has not yet arrived when we can depend solely upon immunization or therapeutics to solve the problem of communicable disease. Where possible these measures should be supplemented by the maintenance of an environment in which communicable disease cannot persis or be transmitted. There are other biological hazards in the environment besides communicable diseases. There are the creatures whose venom is poisonous, e.g., certain snake,s scorpions, black widow spiders, and bees.

Various plants berries, nuts and roots, such as horse chestnuts, toadstools and cator beans are poisonous to man. Individuals may be allergic or acquire an allergy to many foods (eggs, wheat, etc.) to pollens of various plants (ragweed, roses, grasses, trees, etc.), to oils of plants (poison ivy, poison sumac and poison oak), to hair of dogs and cats, house dusts, the bite of insects and other arthropods, and the like. These allergies may manifest themselves by skin eruptions (hives), swelling or other symptoms (similar to respiratory or intestinal diseases).

While not directly related to disease of man, one may want to include in the consideration of biological environmental hazards the economic pests and insects. Pesticides are used to help control unwanted plants, animals and arthropods; however, unexpected and undesirable experiences in effects on wildlife have resulted from the indiscriminate use of pesticies. Research is being directed toward the development and use of natural biological control systems to avoid the problems resulting from chemical agents.

Excessive algal growths have been responsible for fishkills and bad tastes in water. The nuisance of mosquito bites has led to demands for control even where no mosquito borne disease is prevalent. Rats destroy valuable food and present a hazard of rat bites to infants, even when there is norat-borne disease. Perhaps man should be included in the biological hazards to man.

With the decrease in other environmental hazards, some see man's continued high rate of reproduction as a serious threat. In some areas of the world, the increase in population has wiped out the benefits that might have been realized from economic development and the increase in food production.

The rate of increase in the numbers of people on earth began about 200 or 250 years ago to rise above earlier long-term averages. This rage of growth increased even more during the first half of this century and still more after World War II. Some authorities think this rate of growth (1.8% annual average 1950-1965) is nearing its climax. Estimates based on a 1.25% average annual increase forecast a range of about 4.4-7 billion for the year 2000. The population of the world took more than a thousand years to double in the ages prior to 1750. Since then the population approximately doubled in the 150 years to 1900 and again in the years since then. Because of man's destructive habits with respect to other animals and plants,even fellow men, many are concerned about this upsetting of the balance of nature.

Chemical hazards are not new to mankind. Even before the rise of industry some waters were recognized as "bad." The term "miner's consumption" was a recognition that some activities held special peril even before the specific cause was unknown. Man's reaction to a chemical depends upon its toxicity and the individual's susceptibility, the duration of exposure, and the concentration of the chemical. The effect may be related to body weight and the form (solid, liquid, gas) of the chemical at the time of exposure. Occasionally, a slight tolerance may be developed over a period of time from small exposures, e.g. arsenic. Some chemicals are more toxic when the subject is exposed to another chemical simultaneously (synergism). There are hundreds of new chemicals developed each year. For many of these, the toxicity may not be known in the development stages of production, with resulting occupational hazards. Many chemicals in liquid waste effluents find their way to water supplies; chemicals from industrial and auto exhausts are breathed; residues from agricultural pesticides are consumed with food. Even if there should prove to be no demonstrable harm from small increments in each of these elements water, air, food-who can say what the total effect may be over a number of years?

Natural exposures to toxic chemicals may occur where water dissolves the chemical in percolating through the ground. High concentrations of nitrates (methemo-globinemia-"blue babies") and fluoride (mottled tooth enamel) may be acquired in this way. Volcanoes are about the only natural source of toxic gases, unless one considers the gases freed in the mining of coal (mainly methane). Among the toxic substances encountered through breathing are chemically active dust (e.g. lead), gases (e.g., chlorine, carbon monoxide), solvent vapors (mainly hydrocarbons and chlorinated hydrocarbons). Some chemical substances may be absorbed through the skin, particularly solvents.

Where gases or solvent vapors are heavier than air, in addition to any possible toxicity their presence may displace all of the oxygen. Displacement may also occur in sewers, as methane gas is produced in the biological decomposition of the sewage, and sometimes in the cleaning of large tanks or vessels, where vapors of the chemical stored or of the solvent used for cleaning fill the space. If a person enters an atmosphere with zero oxygen, he immediately loses consciousness. A gas mask does not help a rescuer since he requires an air supply. Some gases are irritating at low concentrations,but at higher concentrations hydrogen sulfide may be fatal, and nitrogen dioxide may cause ensilage disease among workers in silos. Allergic reactions may occur after chemical contact depending on individual sensitivity. Chemicals may concentrate in certain issues of plants, animals and fish. As higher forms of life consume lower forms, and the big eat the little, successive concentrations may result in a tenfold increase over the initial concentration. Thus, pesticides washing into a stream may not result in a high concentration at that point, but, through biological concentration, fish or shellfish may become hazardous to man or other life.

Chemicals discharged to the atmosphere may undergo reactions in the presence of moisture or when irradiated by sunlight. Polluted reducing atmospheres, of the New York and London type, have been held responsible for high rates of emphysema and bronchial diseases. The primary pollutants are particulate matter and oxides of sulfur. These occurrences are not well understood at present since the laboratory experiments conducted have failed to reproduce the ill effects of polluted community air at comparable concentrations.

An association between the occurrence of lung cancer and cigarette smoking has been observed, but it has also been noted that urban inhabitants have lung cancer more frequently than rural residents. Some authorities suggest this "urban factor" may be the result of air pollution. Although some chemicals may be hazardous, the proper use of some of these same chemicals can enhance the quality of life. Nitrogen in fertilizer is necessary to the production of crops, but an over-abundance,while increasing the crops yield, will also result in a build-up of nitrates in the ground water. Chlorine is used to destroy pathogens in drinking water and swimming pools and to oxidize wastes but when inhaled is extremely toxic. Too much fluoride in a natural wastes will mottle tooth enable; but when the fluoride content is adjusted to the proper range, tooth decay is prevented and no mottling occurs. While food additives are sometimes questionable, many are beneficial in lengthening shelf life (preservatives) or improving the nutritive qualities (vitamin additions).

Physical hazards in the environment may cause death, disease, or disability. These concerns include a wide range of factors such as dusts, humidity, equipment and environmental design, and radiation. In the previous paragraphs, chemically active dusts were considered. There are other dusts which are not absorbed in the body but cause a fibrosis when deposited in the lungs-silica, asbestos, etc. Some dusts, e.g., cement dust, however, may be a nuisance, but, a far as we know, dc not induce a disease. Temperature and humidity are important to efficient human performance. The critical air temperature for man ranges from 1°C to 32°C and the thermoneutrality zone is 24°-31°C.

Metabolic rate, oxygen consumption, respiration, heart beat, and blood pressure are affected by temperature. Temperature cooling control is not usually specified in housing regulations, although excessive numbers of persons die in summer heat wave. Length of exposure to extremes of temperature and humidity is critical. Electromagnetic radiation includes X-rays, ultraviolet, infrared, and microwaves. Prolonged exposure may induce skin burns, cause elevation in temperature and even death, X-rays cause effects similar to Y-rays as both are ionizing radiations.

Carefully controlled ionizing radiation is invaluable in making radiographic pictures and in therapy; however, excessive exposures may induce cancers, cause genetic mutations and other biological damage. Mechanical vibration and noise can cause hearing loss and produce other psychological and physical disturbances. People living in cities may suffer some hearing loss from community noise. The physical design of the equipment and environment may effect the probability of accidents. Machinery, automobiles, highways, homes, and recreational facilities should be designed to prevent the occurrence of accidents. Accidents are the leading cause of death in most age groups, particularly young adults. Inspite of this fact, only recently has accident control been considered a public health concern. Light, colour, contrast and glare are important in visual tasks from the standpoint of efficiency, fatigue, perception, and comfort.

'Biological systems are sensitive to changes in atmospheric pressure. Furthermore, a rapid decrease of air pressure may result in the release of nitrogen bubbles in the blood stream causing "bends." It is generallv conceded that exercise is essential to the maintenance of a status of health and in the prevention or delay of some degenerative diseases of the middle and later years of life. Thus, the availability of recreational opportunities and facilities that allow for physical exercise is important, particularly in the urban environment.

Sociological and Psychological parameters of environmental design are not as well delineated as biological, chemical and physical factors. A total consideration of the environment of man must consider sociological and Psychological influences (and hazards), even though we may not at this time know their precise effect, the magnitude of their influence, or the epidemiology of mental illness. There is enough information available to suggest that this is an area of serious concern. Dubos asserts that mental characteristics are shaped by the environment, that genes only determine responses to stimuli. Envrionmental stimuli impinge on man from the time of conception. On the other hand, a variety of environmental stimuli are necessary for mental development.

Even though man may adapt to Envrionmental conditions this does not occur without stresses which may ultimately result in disorders of the body and mind. Studies on this subject are difficult to evaluate because of the multiplicity of variables that cannot be isolated to study specific effects. One study found that four out of five inhabitants of a highly congested area had psychiatric disorders, with one in four sufficiently severe to disrupt their daily lives, compared to a low density agrarian community where 97% never experienced a major mental disorder. Noise, overcrowding, lack of privacy, lack of opportunity for social interaction, lack of open space, boredom, compulsory leisure, tedium in travelling from home to work, stresses from work, traffic, crowds, and "estrangement from the conditions and natural cycles under which human evolution took place" are environmental factors which have been identified as having possible social and psychological consequences.

Determination of the sociological and psychological parameters of environmental design will require more definitive research. Findings from studies with animals and some studies of human population indicate that sociological and psychological hazards are just as important as the biological, chemical, and physical factors in the development of positive health and in the prevention disease.

The quality of life is directly related to the quality of the environment. Concepts of "pure" water, "pure" food, "clean" air, "clear" neighborhood reflect the newer concepts of health as meaning more than the absence of disease. the environment must satisfy not only man's physiological needs but also his psychological and sociological needs. The quality of the environment through history has been a measure of the character of a civilization —"a way of life".

3 DAMS

In India 'development' in the upper catchment of rivers has often had an adverse impact both on the ecological balance and the well-being of the local population of the area. Even when an awareness of the ecological and economic problems pertaining to these areas is shown by policy-makers, it suffers from a certain dichotomy which prevents the achievement of the goals of both environmental protection and improvement in the standard of living of the people. This is because "Crude environmentalists" in government circles recommend a package of programmes which will have favourable impact on the ecological balance of these areas without taking into consideration the likely impact of such policies on the economic life of the local population while government departments concerned with improving the standard of living of the people residing in upper catchments of rivers recommend an alternative set of measures to improve the standard of living of these people without paying due regard to the environmental impact of such programmes. Large dams are no more revered unthinkingly as the temples of India's progress. Their destructive consequences are not being increasingly recognized. Even so, bigger and bigger investments are being made in them. All major rivers today are dammed or are in the process of being so. Only a fraction of the country's hydropower potential has been realized. This indicates that in the next two or three decades, dam construction is going to be a major development activity. In the light of recent experience, questions are now being raised about the viability of large dams which have often proved to be ecological disasters. Can there be ecologically sound and economically viable engineering alternatives to large dams? If dams are inevitable for a large country faced with task of feeding its growing millions and for meeting its energy requirements, can the destruction they cause be minimized and adequately managed?

Hardly a day passes without a dignitary taking up cudgels on behalf of the constructing of a big dam somewhere in the country—or against one. If nothing else, this demolishes a myth: that environment is a "non-political" issue. As should be apparent, it is entirely political for the simple reason that it concerns the use of natural resources to which competing classes and communities lay claim. Specifically in relation to "megaproject" like those on the Narmada or at Tehri, there are people who stand to gain : immediately, contractors and their employees, as well as the founders; ultimately, the beneficiaries. In the "opposition" are those who will be displaced, together with environmentalists and human rights activists. While the Sardar Sarovar in Gujarat and Narmada Sagar upstream in MP are probably the biggest among dams being built in the world, and Tehri among the highest, disputes are also simmering over Bodhghat in Bastar, Koel-Karo near Ranchi and Pooyamkutty in Idukki district, among others.

Clearly, the government will have to make up its mind on what its policy towards big dams should be if it does not want to confirm the impression that it is ruling in different directions. This is not a matter of public relations but hard economics. Already, Rs. 400 crores has been spent over 16 years on Tehri and half as much on Sardar Sarovar, every day's delay costs lakhs. Not that money spent should be the clincher: long-term losses, if a project is completed, could outweigh such expenditure. The time-honoured path will be to plump for "development, even while protecting the environment. However, as the heated exchanges between protagonists of big dams (like Mr. Sanat Mehta of the Sardar Sarovar Nigam) and opponents (like Baba Amte) show, the debate has transcended such platitudes. The crux of the matter is : who is benefiting from such projects, and who is paying the cost in social terms as well ? There is no question that big irrigation and power schemes have, since the original "temple of today," Bhakra-Nangal, been an integral part of the agricultural and industrial growth of the country. At the same time, there is no getting away from the fact that for all the euphoria over the Green Revolution, the present annual output of cereals, a total of 170 million tonnes, is on the Planning Commission's own norm of 4 tonnes a hectare from irrigated land—70 million tonnes less than what should be the contribution of irrigated areas alone. As much as 70 per cent of arable land today is rain-fed and even when the total irrigation potential is exhausted, half will always depend on the monsoon. If the present productivity of farming is tremendously low, it is due to poor land and water management.

The two-fold benefits of big dams can be subjected to scrutiny. Take power, which may have camouflage the fact that electricity is not quite the "general good" that it is made out to be, given its present use. Two-thirds of it is consumed by industry. While no one can deny the massive strides made by industry since independence, on one count—employment—its record is abysmal. In 1988, the entire organized sector, which includes both the public and private sector, accounted for only 26 million jobs. Whatever the other benefits of technological advances, increased employment is certainly not one of them. Factories, between them, hired less than 8 million hands in 1987. Five-year plans have allocated one-fifth of the resources to the power sector alone and this is considered a *sine quo non* of progress.

Unfortunately, planners mistake electricity for energy. For that matter, only "commercial" energy is accounted for, the consumed in the country is by way of fuelwood for cooking, and most of it is collected "free"—that is, at great expense of human labour. If one looks at the needs of the neediest, therefore, it is futile pointing to electricity.

Only Luddites would argue against the need to generate power. As energy experts like Dr. Amulya Reddy have long observed people's homes should be the first to be electrified; in practice, rural electrification is the bane of all state electricity boards. The social and economic spin-offs of extending the working day would be for greater than those of lopsides industrial growth. Even though the country has been able to electrify nearly eight out of every ten villages, only, 14 per cent of the households in a typical village on the grid can afford to install connections to their homes. This is the context the debate over big dams.

The opponents of big dams are'also often barking up the wrong tree. It is the entire development process that is being debated, not but a "big versus small" slanging match. If the policies followed in the last 20 years of encouraging capital and resource-intensive inputs in agriculture are to be continue, together with a highly selective pattern of industrialization (mirrored in uneven urbanization), there is no way of reducing the nation's dependence on such projects.

Meanwhile, Chipko Movement leader Chandi Prasad Bhatt has started a campaign against the Vishnuprayag Hydroelectric Project because of the dangers it poses to the unique ecosystem of the Valley of flowers and the general geological havoc it could create. Bhatt, as he has shown with the Chipko Movement, is capable of mobilizing strong support among the local population though, a section of the workers who have found jobs on the project oppose the campaign. Another instance of intervention at the highest levels resulting of a large dam is Lalpur in Gujarat. Gadhian social worker, Harivallabh Parikh, an ardent supporter of Mrs. Gandhi and a close associate of Jinabhai Darji, an influential Congress (I) leader in the state runs an ashram at Rangpur, which would have been submerged if the project had materialized. He raised two issues—destruction of forests and displacement of tribals. A committee is preparing an alternative plan.

These very issues are cited by Baba Amte, while mobilizing protest rallies against the Sardar Sarovar, Bhopalpatnam and Inchampalli Dams. He is also questioning the very raison d'etre big dams by asking : whose development do these dams represent? Amte's strategy is to mobilize people through mass rallies and public opinion by getting in touch with environmentalists, political leaders, planners, journalists and even the Prime Minister himself through meetings, discussions and personal letters. The protest against the Tehri dam in Garwal is another, where along with rehabilitation, environmental issues relating to large dams have also been highlighted.

The Bandh Virodhi Sangharsh Samiti led by advocate V.D. Saklani has underlined the threat to the ecological balance of this sensitive mountain region through rallies, seminars, and even wall writing. The movement has gained momentum and local support because of the strident protests by citizens of Tehri town, which will be submerged by the hydro project.

Tribal population who are affected by a large dam are fully aware that they will lose their trees, land and houses. The Munda tribals of Chhota Nagpur likely to be affected by the Koel Karo Project are well organized, with a coordination committee and a representative from each village. Local leaders of various political parties are also part of the organization, making it probably the strongest popular movement in India against a dam. The movement has a single demand : comprehensive rehabilitation—social, cultural and economic.

The Koel Karo Jana Sangathan is involved in negotiations with the government for getting as much compensation as possible. In Koel Karo, people succeeded in stopping work by their Kam Roko (Stop Work) Movement in 1978 and want the authorities to involve the affected people in fixing rehabilitation terms. In mid-1984, the new Chief Minister of the state had ordered armed police into the area to put down the tribals. By contrast, in the absence of organized protests or support from any voluntary groups, people's houses were bulldozed for the Sarisailam Project in Andhra Pradesh and the villagers were hounded out by demolition squads.

As far as rehabilitation is concerned the movement in the villages affected by the Sardar Sarovar dam on the Narmada is setting an example. The Narmada Tribunal Award spelt out terms for villagers displaced from Madhya Pradesh and Maharashtra and, under pressure from various voluntary groups, the Gujarat government had to agree to similar terms. To ensure that the rehabilitation programme is carried out properly volunteers of the Gujarat Chhatra Yuva Sangharsh Vahini have gone into the villages and are both educating an organizing the villagers. Researchers from the Centre for social studies at Surat have conducted a detailed house-to-house survey in all the villages and presented detailed reports on their needs. At the state level, there are other voluntary groups which have been instrumental in bringing pressure to bear on the government.

Large dams are generally being constructed in areas which have so far not been touched by modern development. Many people in these areas look at such projects as symbols of hope. They even accept the argument that someone has to pay the cost for the benefits to society, even if they are the ones who have to pay. As things are today, people are not likely to oppose dams en masse. But they are definitely getting ready to agitate for better rehabilitation terms. But also as more and more dams are chosen on environmentally sensitive sites, as in the Himalayas, or in heavily forested areas, protests against dams are likely to grow.

Meanwhile, a number of experts have even begun to question the very usefulness of the large surface irrigation systems that dams promised, which means that a thorough analysis of India's current water management plans - and especially of dams—is urgently needed. In most cases, as in Koel Karo, these groups are fighting mainly for better terms of rehabilitation. But in a few cases, citizens' groups have even begun to demand that a dam or a hydroelectric project not be built at all either because of the number of people it would displace, as with Bedthi or with Bhopalpatnam and Inchampalli, or because of the unique ecosystem it would destroy, as in the case of Silent Valley or the Vishnuprayag Hydroelectric Project.

Some of the antidam campaigns have even been successful in stopping these massive projects : Silent valley and Bedthi have already been given up by the

Kerala and Karnataka governments respectively. And government committees are investigating the environmental impact of the Lalpur dam, the Vishnuprayag hydroelectric project and the Bhopalpatnam and Inchampalli dams.

A large dam inevitably alters the course of nature and, because of the play of complex and largely unknown forces, the ecological consequences are usually much more serious than were bargained. Silent Valley has been dropped but the government has decided to go ahead with Tehri, Vishnuprayag and many other projects. Also being proposed is the Manibhadra dam in Orissa, in one of the prime forest areas remaining in the country. The forest cover bears the brunt of dam construction in more than one way. In the Himalayan valleys, contractors who come to clear the forest which is going to be submerged greedily cut trees even in unaffected areas.

People who are being displaced from Tehri town in Garhwal to make way for the dam will be rehabilitated on forest land that has been acquired and cleared near Hardwar and Rishikesh. The forest is also being cleared for approach roads, offices, residential quarters and for storage of construction material. With the reduction in forest cover and the entry of people the pressure on the remaining forest increases. Their need for firewood leads to further denudation.

The construction of a dam, therefore, has a multiplier effect. The 1983-84 annual report of the Department of Environment reveals that the construction of the Idukki Dam over the Periyar river in Kerala has hastened the degradation of vegetation and sharply reduced the forest cover. It is estimated that during the period 1951-1976, 0.49 million hectares of forest has been lost due to major river valley projects. This is roughly 10 per cent of the area that came under canal irrigation—5.5 million hectares.

Despite the increased awareness about the need to protect forest cover, little attention is paid to this factor when a site is selected for a dam, basically because all forests are state property and no one has to be compensated for them. Even in cost benefit calculations, forest loss is grossly undervalued and the recurring loss in terms of timber and firewood yield per year and social costs of degradation of environment are never taken into account.

Under valuation of standing forest helps in boosting the cost-benefit ratio for obtaining Planning Commission approval and finance from international agencies like the World Bank. Moreover, the contract for clearing the forest is also undervalued thus leaving a bigger margin to be shared between the contractors and the forest department. Because of public protest, some provision is now made for reforestation or compensatory afforestation. In Tehri, for instance, Rs. 10 crore have been provided for this. With the best will in the world, it will take at least 10 to 12 years for forest cover to appear. Compensatory afforestation in catchment areas should, therefore, start with the planning of the project, since it will also help in reducing flow of silt into the reservoir.

It is unfortunate that, despite the massive dams built over three decades, there is no rehabilitation policy worth the name. Rehabilitation becomes much more important and difficult when most of the displaced are tribals. Tribal life and culture is so uniquely interwoven and integrated with the forests that once uprooted, they are totally disoriented. The main problems regarding rehabilitation have been well evoked by the remarkable studies on the gigantic Sardar Sarovar (Narmada) Project in Gujarat conducted by the Centre of Social Studies (CSS) in Surat and the Lokayan study on Sarisailam. The issues can be summarized as follows: rehabilitation has always been considered an obstacle in implementing a project and an extra financial burden; then authorities always try and get away with minimum; no guidelines exist to include even the barest arrangements for rehabilitation in project plans; and facilities provided differ from project to project. Provision of measures like "land for land" and basic amenities like drinking water, schools, dispensaries and so on were made obligatory in the case of Sardar Sarovar. That had been made possible because of a dispute between the three states—Maharashtra, Madhya Pradesh and Gujarat—which was handed over to a tribunal.

The tribunal outlined the rehabilitation arrangement to be carried out by the Gujarat government. As Gujarat was to bear the costs of rehabilitation of the affected villagers of Maharashtra and Madhya Pradesh these two states saw to it that fairly elaborate measures were given in the tribunal's award. The award was silent about the rehabilitation of 'oustees' from affected villages in Gujarat. It was only after a tough stand by voluntary agencies that the Gujarat government agreed to similar provisions for its own oustees. When Gujarat got down to implementing these measures, it was reported to be shirking its promises. In any event, the Narmada Award has become a precedent for subsequent hydroprojects.

The only legal power with the government to handle rehabilitation is Land Acquisition Act, enacted by the British in 1894. Apart from cash compensation, any effort to resettle people is considered charity. Compensation for land and houses is usually grossly inadequate, and none has ever been paid for loss of employment or disruption of livelihood other than from land. Landless labourers, marginal farmers, sharecroppers, tenant cultivators and those who live on collecting and selling forest produce have suffered most. Village artisans whose base is the village also get uprooted. Small farmers who supplement their income by rearing cattle are hit because new settlements never have enough grazing land. Villagers are forced to sell off their cattle at throwaway prices before shifting. People engaged in non-agricultural activity, like fishing are also not entitled to compensation. For example, there are 750 families along a 150 km river stretch downstream of Broach town who make a living out of catching hilsa fish. The construction of Sardar Sarovar will interfere with hilsa migration and 750 families will be rendered unemployed. No one in authority even pauses to ponder over the fate of the people involved in so-called marginal occupations. To add insult to injury, the meagre rehabilitation measures are implemented half-heartedly. The entire process, beginning with the notices issued to the villagers for acquisition of land to the final settlement, is a story of misery. Poor people are harassed, hounded out of their villages and handicapped for the rest of their lives. The whole operation is conducted in a most unimaginative and bureaucratic manner as the story of oustees from the Sarisailam Project demonstrates.

The law requires that written notice be issued individually to the villagers for acquisition of their land and houses. The villagers, mostly tribals, are illiterate and ignorant of the legalities and get confused. As the experience of the Koel-Karo project in Bihar shows, panic spreads from much earlier when the project is being planned.

Rumours abound about the rate of compensation and resettlement sites. In the case of Sarisailam, the notice specified terms of compensation with a clause that appeals about the rate that had to be submitted in writing by stipulated date. The rate of compensation was very low and depended upon the type of land, house and so on thus leaving a lot of elbow room to the land acquisition officers. Though the law provided for compensation at the market rate according to the latest land deals in the area, the compensation was arbitrarily fixed by the authorities. It has been the experience everywhere that the rich farmers bribe the officers and manage to get higher compensation and better alternative lands, whereas the poor keep running from pillar to post.

Vidut Joshi of the CSS in Surat points out that the terms of acquisition and other procedures were never orally explained to the people in the Sardar Sarovar project and in the process illiterate people were cheated by the officers and their intermediaries, except in cases where activists of voluntary organizations come to their help. Even when compensation is paid to villagers through their bank accounts, it has been found that part of it goes in sustaining them during the shift to a new site and much of the rest is frittered away in non-productive expenditure.

Another problem is that government takes only that land which is to be submerged or is required for other purposes, leaving many partly affected people in the lurch. When part of a village is shifted elsewhere, the families left behind face a life of isolation, in some cases, when the major portion of a family's land is taken over, it can neither survive on the remaining area nor get any compensation for the remaining land or for its house if it decides to shift along with fellow villagers.

Such cases were never considered until the Narmada Tribunal specified that if 75 per cent of a family's land was affected, the entire land should be taken over and it should also be paid compensation for the house and transport for shifting. Even now, there is no provision for families who are isolated in partly affected villages. Almost all the farmers left in the vicinity of Shivsagar (Koyna) reservoir live in scattered basis, below the poverty line as they have no means of subsistence except for a few quintals of paddy grown by primitive methods.

"Land for land" is a better policy than cash settlement and the Narmada Tribunal has in fact specified that a minimum of 5 acres should be given to each affected family. Even in implementing this policy, the land is not given in the command area in most cases, forest land is either cleared on waste fallow land given without any provision for developing the land or for the supply of necessary inputs; a village is broken up and families dispersed; villagers are usually left to buy private land, take loans from the government, which puts poor villagers at a disadvantage—land prices in neighbouring villages shoot up steeply if the government takes up resettlement, the villagers are resettled in distant places, sometimes in a totally alien environment and culture, thus creating insurmountable adjustment problems. Oustees from Pong dam in Himachal Pradesh were settled in Anupgarh in Rajasthan, bordering on Pakistan. Though the land falls in the command area of the Rajasthan canal the villagers, who were used to hill cultivation, found it difficult to adjust to the rigours of the desert.

Tribals affected by Inchampalli and Bhopalpatnam in Maharashtra do not even speak Marathi and if they are resettled away from forests and in an alien Marathi speaking area they will find it difficult to survive and shifting them to any other area would amount to 'ethnocide,' in the words of Baba Amte, the social activist who has been leading a mass movement against the two dams. When the construction schedule of a dam is planned, residential quarters for officers and other staff are always given priority. In the Sardar Sarovar project, Kevadia colony; a well-developed residential area with markets, schools, parks and other facilities was kept ready before the staff moved in. On the other hand, when villagers were served legal notices for acquisition of land, no preparation was made for their shift to alternative sites. Even if some sites were located, no effort was made to develop the sites. The people were generally left to fend for themselves. Arrangements for drinking water, dispensaries, schools, village roads or drainage of the rehabilitation sites are only completed years later.

In the case of the Ukai Dam in Gujarat, resettlement work was undertaken by the Ukai Nav Nirman Samity, a voluntary agency. What is more, the then Chief Minister of Gujarat belonged to the same district. Even so, out of a total of 18,500 affected families, only 3500 families could be resettled.

Though Rs. 7 crores was paid as compensation, only Rs. one crore was deposited in banks, the rest being squandered or used for day-to-day needs. In the case of the Koyna Dam, when a survey of displaced people was conducted by Deccan College in Pune in 1969, it was found that around 12 families from practically every village had returned to their neighbourhood. Since those who are displaced are the illiterate and unorganized rural poor, they do not have any political clout. Whenever they have been organized with outside help they have succeeded in getting a better deal.

The Ukai Nav Nirman Samiti handled the rehabilitation with some success with the support of the then Chief Minister of Gujarat, Baba Amte in Gadchiroli District, Maharashtra and the Koel-Karo Jana Sangathan near Ranchi, are raising issues which are fundamental to a rehabilitation policy.

The Koel Karo people are demanding that the government spell out comprehensive rehabilitation policy—economic, social and cultural to the satisfaction of the people affected by the project. The struggle committee consists of heads of all the villages involved and they have succeeded in stalling the construction of even the approach roads to the dam site. The state government has been forced to enter into negotiations with the people's representatives. The proposal put forward by the state government includes reservations for the displaced people in Class III and Class IV jobs and setting up of vocational training facilities for the youth but provision of land-for-land is not part of the deal. In Gujarat, four voluntary agencies have been involved in rehabilitating the villagers displaced by Sardar Sarovar on the Narmada. Chhatra Yuva Sangharsh Vahini volunteers have gone into the villages to educate villagers about compensation and rehabilitation and mobilize them against corruption and for a better deal in accordance with the Narmada Tribunal Award.

The Action Research in Community Health and Development Group of Mangrol and the Rajpipla Ṣocial Service Society are working in cooperation with the Sangharsh Vahini and are liaising between the people and state government. The CCS in Surat has done commendable work in studying each affected village in detail, and indicating the rehabilitation requirements of the people.

A comprehensive rehabilitation policy can be spelt out only when development is understood as a process which benefits the people. If the completion of a project at the minimum cost becomes the overriding objective, rehabilitation simply amounts to evacuation. People who could previously barely manage to survive in their traditional environment are uprooted as a result. If people are the major focus, any project should seed to improve the conditions of those living in object poverty.

According to the CCS, the objectives of rehabilitation should be :

- The people displaced should get an appropriate share in the fruits of development.

- They should be rehabilitated by creating new settlements with their own environment.
- Removal of poverty should also be an objective of the rehabilitation policy and therefor some land to all.
- Oustees (even the landless) should be given assurance of employment.
- While dealing with tribals one should also keep in mind the following five principles of tribal-development accepted during Jawaharlal Nehru's era as 'tribal panchsheel.'
- Tribal should develop along the lines of their own genius and we should avoid imposing anything on them.
- We should try to encourage their own traditional arts and culture in every way.
- Resettlement should be in the neighbourhood of their own environment.

  If resettlement is not possible in the command area, top priority should be given to the development of irrigation facilities and supply of basic inputs for agriculture; drinking water, wells, grazing grounds for cattle, schools for the children, primary health care units and other amenities should be arranged.
- In partly affected village, villagers should be given the option of shifting out with others with the same compensation as available to evacuees.
- Training facilities should be set up to upgrade the skills of affected people and reservation in jobs should be made for the willing adults among the evacuees.

- Special attention should be given to the rehabilitation of artisans and village crafts people.
- Villagers should be taken into confidence at every stage of implementation and they should be educated, through open meetings and discussion, about the legalities of the Land Acquisition Act and other rehabilitation provisions.
- The aid of voluntary agencies should be taken in the planning and implementation of the rehabilitation programme.

It can be argued that it is inevitable that any development project will benefit some citizens more than others and invariably some citizens may have to bear the additional cost. In the case of big dams, like most other development projects, the direct beneficiaries are generally the rural rich and the educated urban elites whereas the direct social costs are borne by the rural poor, mainly tribals. If people who pay the social costs start to count in the development strategy, the entire orientation of the planning process will have to change.

Storage of water by building large dams is regarded by irrigation engineers as a crucial component of India's water management system. The Central Water Commission argues that benefits like hydroelectricity, irrigation, flood control, industrial and municipal water supply are not possible "unless the water is stored in reservoirs created by dams."

Water storage is necessary because rainfall is very unevenly distributed, varying from 10,000 mm in Cherapunji to about 100 mm in western parts of Rajasthan. There are also considerable seasonal variations. Over 75 per cent of the rainfall is received during the monsoon period of June to October and even during this period there are many erratic spells of heavy rain and dry weather. On one hand as many as 99 districts are classified as drought prone; on the other, 85 per cent to 99 per cent of the monsoon water flows as waste to the sea. Since independence, about 700 storage dams have been constructed. The total storage capacity created so far, including those of small tanks, is assessed to be about 16 million hectare-metre. Out of a total of 178 mham surface water available in the country, only 27 mham are now being used and the utilisable potential is about 67 mham. About 90 per cent of the water being used is for irrigation alone.

By the year 2000, the government hopes to provide irrigation to 113 million hectares, the ultimate irrigation potential out of a gross cropped area of 200 million hectares. Major and medium irrigation projects would irrigate about 584 million hectares and minor irrigation projects 54.6 million hectares. Between 1980 and 2000, some Rs. 60,000 crore (at 1980 prices) will be required for the development of this irrigation potential. At the beginning of the Sixth Plan, a total potential of 56.51 million hectares had been created of which 26.61 million hectares was through major and medium schemes and the rest through minor irrigation. By June 1984, about 65.6 million hectares of irrigation potential had been created.

As even by the year 2000 nearly 40 per cent of cropped area will be unirrigated. The government is preparing a national perspective plan for water resources development, which will allow for large scale inter-basin transfer of water from water surplus river basins to water scarce river basins. This will mean that an additional 22 mham can be utilized. But these inter basin transfers of water will again depend primarily on the creation of storage reservoirs: the ultimate storage capacity in that case has been assessed at 37.5 mham. The irrigation potential will thus increase from 113 million hectares to 148 million hectares with about 25 million hectares coming under surface schemes and about 10 million hectares benefiting from increased use of ground water. This ambitious plan will require an investment of about Rs. 50,000 crore. The investigation and planning work of the National Water Development Agency charged with the development of the perspective plan will cost Rs. 107 crores spread over 7 to 10 years. The Commission argues that if water resources have to be harnessed by building storage reservoirs then either developed areas with a large population, agricultural areas of forest areas with a tribal population, have to be submerged :..." If none of these are to be affected... we may have to stop all further progress in water resources development.

Thus, if government programmes go ahead as scheduled, there will be hardly any free-flowing rivers left in the country. Already the Cauvery is fully exploited. The Krishna and Godavari rivers, with large dams like Tungbhadra. Srisailam, Nagarjunasagar, Mula, Jayakwari completed or near completion and several new rivers in the offing like Inchampalli and Polavaram should be fully exploited soon.

With the award of the Narmada Tribunal, the Narmada valley development programme has also been launched and by the early part of the next century, the Narmada will be a fully controlled river. The Tapti with Ukai and Upper Tapti dams will soon become a fully used river. Mahi with Bajajsagar and Kadana dams faces the same prospect, and so does the Betwa with Paricha, Dhukwan, Orchha, Matatila and Rajghat dams. The waters of the Sone with Bamraha Bansagar, Sukhar Churhat, Hurma, Rihand and Kanhar will have been bottled up by 1995. The Mahanadi with Dudhawa, Ravishankarsagar, Hasdeo, Jonk, Arpa, Sheonath, Pairi, Hirakund and Tikkarpara dams will have been mostly used up in the next 20 years. The rivers in the Indus Basin, Sutlej, Beas and Ravi are also being controlled and the northern tributaries of the Ganga (Bhagirathi, Sharda, Ghagra and Kosi) too are likely to be fully exploited by the turn of the century.

But whatever the Central Water Commission may argue, more and more experts are beginning to question the very utility of big dams, and especially of the large surface irrigation schemes they offer to the country. In a study on canal irrigation and its impact on crop production and environment, D.R. Bhumbla, former Agriculture Commissioner and Vice Chancellor of Haryana Agricultural University, argues that "from the experience of major and medium irrigation works in India it is evident that its benefits in arid areas though spectacular for the first 10 to 20 years, gradually get reduced and a considerable portion of the land gets deteriorated because of water logging and salinity.

In the humid areas, the benefits are doubtful from the beginning and in many cases negative. The programmes of major and medium irrigation works as has been envisaged for future in humid areas, in my opinion, would not only lead to disastrous consequences in degradation of soil and environment but would also result in reduced agricultural production. It is time to halt the expansion of these programmes..." In low rainfall areas of Rajasthan and similar areas of Punjab and Haryana (300 mm to 400 mm), canal irrigation is necessary because raising crops without irrigation is either not possible or uneconomical.

Moreover, the ground water is saline in many areas and hence the potential for tubewell irrgation is limited. The introduction of canals in these areas has resulted in a significant increase in yields of crops like cotton in the kharif season and wheat in the rabi. In semi-arid regions of Tamil Nadu, Andhra Pradesh and Karnataka, where rice and sugarcane are the main irrigated crops, increases in yields have been significant. But in all these areas water-logging and soil salinisation have emerged as serious problems. As far as unlined canals and distriutaries are concerned, a study by the Central Water and Power Commission in 1967 revealed that about 71 per cent of the water is lost in transit from the reservoir to the field as highlighted in Table 1.

Table 1

Loss of water from unlined canals due to seepage (percent)

Canals	15
Distributaries	7
Watercourses	22
Field Losses	27
Total Loss	71

The water table in the Bhakra canal command area has been rising at the rate of about one metre per year. The managing director of Haryana State Minor Irrigation and Tubewell Corporation, in a press conference in late 1983, expressed serious concern over the rising ground water table in the state. He said that 65 per cent of the area in the state covered by canal commands had saline ground water aquifers underneath and in 14 lakh hectares, the water table had entered the critical zone of upto 3 metres depth. This was creating water logging accompanied by salinisation. The rising water table will affect the production of many crops adversely, the story of canal irrigated areas of Pakistan where cotton has vanished from many fields, is likely to be repeated, says Bhumbla. The high water table, apart from affecting agricultural productivity, has many other adverse effects. Because of the loss of moisture storing capacity of the soil, even slight rains cause floods. The maintenance of road and buildings becomes difficult and diseases increase because of water-logging. In Hissar in Haryana, the load bearing strength of the soil has become less than 50 per cent in 15 years. In the sub-humid and humid areas of the country (rainfall more than 1000 mm), Bhumbla argues that the benefits of canal irrigation are questionable even in the short run. In fact, in many areas, introduction of canals has resulted in reduction in productivity of crops. In Bihar, Madhya Pradesh and Orissa, which together have 2.5 million hectares of canal irrigated area and where rice is the main irrigated crop, the yield per hectare is not only less than one tonne, but there has been perceptible increase for the last 20 years. In Uttar Pradesh, which has the maximum canal irrigated area apart, there is a general complaint that from pre-sowing irrigation, hardly one irrigation is possible and the wheat crop frequently suffers from moisture stress.

During the drought of 1979-80, the farmers in the Sarda Sahayak Command Area complained there was no water in the canal but in the wet year of 1978, the canal water aggravated the flood problem created by excessive rain. In the rice growing area, water is seldom available for raising rice nurseries before the start of the monsoon rains. Thus, in the non-rainy season, winter or summer, when irrigation could result in significant increases in yield enough water is not available. Further, in the humid areas, it becomes difficult to message canal water during the rainy season and the problem of water logging becomes serious in a much shorter time. In fact, this may aggravate the problem of drainage and result in loss of production. The situation is likely to be much worse as canal irrigation is introduced in high rainfall areas particularly if the soils have low permeability as has been demonstrated by the situation in Tawa and Barna command areas, where yields of all crops have declined since the introduction of canals. Bhumbia therefore concludes : The effect of canal irrigation on increasing cropping intensity is marginal." He argues that the irrigation departments do not even produce relevant data to show the cropping intensity inlands irrigated by canals and tubewells. All that we know is that in 1976-77, the net irrigate area was 34.8 million hectares, and the gross irrigated area was 43.1 million hectares, which gives a cropping intensity of 1.24 which is very low and a good reason for the low productivity of irrigated lands in India. Bhumbla argues that in most canal irrigated areas cropping intensity is less than that in tubewell irrigated areas, generally two or more. In other words, a substantial increase in food production in the last decade or so has actually come from tubewell irrigated areas. Bhumbla argues that the government should concentrate on ground water development and management of water by storing it in catchment areas in small reservoirs of canal irrigation schemes.

Ground water development, minor irrigation schemes and micro hydel project are often cited as alternative to big dams, but are looked down upon by power engineers and planners. Small schemes cannot be considered as alternative to big dams for water management and power generation in all cases. But there is definitely a strong case for ground water development and minor scheme for irrigation, while power generation needs can be partly met through small hydel projects. Like surface water, ground water resources in the country are being depleted because of the lower water retention in catchment areas which have been deprived of forest cover. The development of catchment areas, therefore, should become an integral part of an irrigation programme. This will also check the flow of silt into big reservoirs. There have been calls for the creation of a catchment area development authority.

Table 2
Net irrigated area by sources
(million hectares)

	Canals	Tubewells	Total
1950-51	83	NA	20.9
1960-61	10.4	0.13	24.7
1970-71	12.9	4.46	31.1
1978-79	150	8.2	38.0

Wherever water resources have been harnessed through small reservoirs by construction bunds or small dams, the lives and environment of the people in the area have been transformed without anyone having to pay the cost of development. Whether it is through the voluntary efforts of Prembhai, recipient of the Jamnalal Bajaj Award for Rural Development, who has been instrumental in the construction of over 750 small dams in Mirzapur district of Uttar Pradesh or the initiative of P.R. Mishra, a soil conservation expert at Sukhomajri near Chandigarh, small dams have generated far more benefits than large irrigation schemes in terms of utilization of water, improvement of agricultural productivity and ecological enhancement.

The Tehri dam, if it comes into existence, will be one of the highest dams in the world and will harness the water flows of two important Himalayan rivers, the Bhagirathi and the Bhilangana. It will also take advantage of the sheer drops in height at many points along the river course for generating hydel electricity.

It is part of a larger plan formulated for tapping the discharge of the Upper Ganga basin wherein some 7400 million cubic metres of water are said to be going waste. This larger plan envisages three storage schemes viz, a reservoir on the Bhagirathi at Tehri, another at Kotli Bhel a little downstream,and a third at Utyasu on the other tributary of the Ganga, the Alkananda. The underlying assumption of the larger plan was that these three large-scale dams would provide benefits to society far in excess of the investment necessary for the project. Over the years however,the validity of this assumption has been reduced due to experiences on the other major water resources development projects in the country, which have not yielded benefits according to expectations. The controversy, however, centres around an important physical feature of the Tehri location, seismicity. It is feared that if an earthquake of an collapse,and the catastrophe would lead to consequences which would greatly outweigh the benefits from development. The project was conceived in 1949 by the Geological Survey of India. The Tehri site was considered suitable provided sub-surface investigations did not reveal any adverse factors.

However, it was only in 1963 that detailed investigations were made. After the visit of the then Union Minister for Irrigation and Power, Shri K.L. Rao, in 1965, the site was finally confirmed. By 1967, several Indian and foreign experts visited the dam site and recorded their opinions. The cost of the project, when first prepared in 1967 was put at Rs. 126.8 crores, in 1989 to Rs. 3,000 crores and in March 1990 to Rs. 4,142 crores. However, Rs. 450 crores is said to have been spent on the project.

Apprehension was expressed with regard to the geologically vulnerable area at around the site, the unstable hill slopes which would constitute the reservoir rim areas and the seismic danger at the site. It was, therefore, decided that it could be a clay core rock - fill dam which would allow more flexibility in its structure. The preliminary investigations by the Geological Survey of India had revealed a river bed fault at the dam site and this strengthened the decision to opt for a concrete dam. The Tehri dam which is finally expected to be 260

.5m high, will be fifth highest dam in the world. It will impound 3.22 million cubic m of water. The reservoir will extend upto 45 kms in the Bhagirathi valley and 25 kms in the Bhilangana valley with a water spread area of 42.5 sq. kms.

The local people have been opposed to the dam from its very inception. They consider the Bhagirathi a sacred river and they fear that a number of holy places downstream would be irreparably harmed because of the dam. The people also have a fierce attachment to their land and hence they have not taken kindly to the plant to uproot them and tear their whole existence apart. The people are also aware of the poor record of the Government in rehabilitating the dam oustees. For example, in the case of Ukai Dam in Gujarat, resettlement work was undertaken by the Ukai Nav Nirman Samiti, a voluntary agency. The Chief Minister of Gujarat belonged to the same district. Even so, out of total of 18,500 affected families only 3,500 families could be resettled. They also fear that a big dam in their area, because of its complex geological and seismic conditions will fail sooner or later and inundate the entire valley washing away everything the hold dear and sacred. This fear has been so widespread in the Bhagirathi valley that the people have spontaneously rallied round to oppose the dam.

In 1965 when the Union Minister Shri K.L. Rao visited the area, he met the representatives of the local people who explained their opposition to the project. In 1973 the local unit of the Communist Party of India organised an agitation for the same cause. In the meantime, the engineers working on the project prepared a Detailed Project Report in 1969 and submitted it to the Planning Commission. The report envisaged the generation of only 600 MW of power at this stage. It was approved by the Planning Commission in 1972.

Administrative approval of the Uttar Pradesh Government was obtained in 1976. Thereafter thirtyfive Gaon Sabhas in the Tehri district passed a joint resolution opposing the project. The Zilla Parishad called upon the Government to drop the project. Inspite of this the Uttar Pradesh Government decided to go ahead with it. The people then decided on concerted action to exert pressure on the Government. The Tehri Bandh Virodhi Sangharsh Samiti, TBVSS (Anti Tehri Dam Committee), thus came into existence. All the national political parties including the Congress (I) and, of course, the local people have their representatives on TBVSS. The first resolution of the TBVSS vehemently opposing the dam, was passed unanimously in January 1978.

Despite this, in March 1978 the Uttar Pradesh Government awarded the contracts for the construction of diversion tunnels. This led to mass protests and a 'satyagraha' in which even old women also took part. In August 1978 a huge mass of silt fell into the Bhagirathi river at Kanaudiagad. It blocked the river flow and created an artificial lake which later broke upon, causing a flood and destruction of property in many villages down-stream. Then a delegation of TBVSS met the then Prime Minister Shri Morarji Desai, and suggested an alternative site for the dam near Deoprayag. This site meant a much smaller area of submergence. Apparently no action on this suggestion.

However, the then Prime Minister sensed the distress of the local people and asked the Department of Science and Technology to assess the environmental impact of the dam. Accordingly, a 'Working Group for the Assessment of the Environmental Impact of Tehri Dam' headed by Shri S.K. Seth was formed in December 1979. In February 1980 Shri S.K. Roy became the Chairman of the Working Group and an interim report was submitted in May 1980.

The Working Group observed that enough scientific data was not available for reaching definite conclusions on the controversial aspects, particularly the geological and seismic consequences of the dam. The Working Group, therefore, asked the Wadia Institute of Himalayan Geology, Dehra Dun, to investigate the 'slope-stability' of the rim area and the catchment area of the dam. The National Geophysical Research Institute (NGRI), Hyderabad, was assigned the task of investigating the impact of Marh Fault name of a geological fault in the area. The preliminary investigations had inferred that this fault lay about 7.5 kms below the dam site. This would have made the dam very risky.

And Roorkee University was asked to conduct a study on the gravity lineament tectonics.

The Working Group also specified that :

(a) Continuous monitoring of the faults lying within a radius of 100 kms of the Tehri dam must be made at numerous locations.

(b) Seismic monitoring had to be done continually group a Statistical data base and to take appropriate control measures.

(c) A detailed geomorphological map of the catchment had to be prepared.

The Working Group also asked for fresh siltation-rate studies to be carried out at number of measuring stations on the river upstream of the proposed dam site. During one of its site visits the Working Group observed : "There was disturbing evidence of what can best be described as an endangered catchment area. Overgrazing and deforestation and land use practices requiring maximisation of the cultivable land even on some precipitous hill sides have created lasting crisis conditions. The obviously unstable hill sides,the landslides visible at all points of road building activity, and the blockage of the Bhagirathi at Kanaudia are indicative of future danger."

The other recommendations of the Working Group included the following :

(a) Incorporation of adequate flood cushion capacity in the dam design,

(b) declaring a major portion of the catchment area as National Reserve land,

(c) ensuring that adequate steps are taken for afforestation and soil conservation,

(d) stabilisation of land-slips, land-slides and other slip-zones,

(e) banning all commercial forestry in the catchment area,

(f) coordination of all forestry to develop the catchment area through a Catchment Area Development Authority and a new organisation called Conservation Corps.

In the end, however, the Working Group observed that it did not have sufficient data to ask on the dam to be stopped. Work on the dam began in 1977 and continued steadily. The project kept making substantial changes in their plans. Thus, in 1978, the initial proposal to install 600 MW of hydel power was revised to 100 MW. Then, in 1983 when proposal was put up before the Soviet authorities for financial assistance, the installed capacity of an other dam at Koteshwar.

The TBVSS continued to make representations to both the UP State Government and the Union Government against the project, pointing out the gigantic risk involved in it. In August 1980 TBVSS President Shri

V.D. Saklani, an eminent freedom fighter and a lawyer from Tehri requested the Planning Commission to review the Tehri project once again. The Commission responded by saying that the Working Group would look into all aspects of the matter. Since work on the dam had already started the TBVSS suggested an alternative scheme to make use of the expenditure already incurred by the project authority.

In this scheme TBVSS suggested that one of the right bank tunnels already constructed might be extended up to Deoprayag down stream and that the resultant drop in height be utilised for generating power by using the tunnel cum-channels method. TBVSS representatives during this time continued to attend the meetings of the Working Group. They (as TBVSS was to allege in 1986 in its writ petition before the Supreme Court) "gained the impression that the engineers' lobby, which was wholly at cross purposes with nature, was playing a game of hide and seek and was dominating the Working Group". In the tug-of-war between the engineer members and the environment-conscious members of the Working Group" no conclusions were reached. The Group was, therefore, not able to submit its final report.

On 20 November 1985 TBVSS filed a writ petition in the Supreme Court challenging the soundness of the plan to construct the dam at Tehri. In the writ petition TBVSS asserted that:

(a) the dam to be located in a seismically active and tectonically risky area had a high probability of failure and would therefore endanger the lives and property of people at settlements such as Munikireti, Rishikesh, Haridwar among others situated on the Ganga downstream,

(b) habitations like Uttarkashi situated on the river Bhagirathi upstream of the inlet of the reservoir from Dharasu would be lost, and

(c) it would create a feeling of insecurity, among the people at the various locations on the reservoir rim slopes and therefore the project would jeopardise the "right-to life" guaranteed to citizens under the Constitution.

The Tehri dam case, TBVSS v the State of Uttar Pradesh (1986) and TBVSS v Union of India and others, is currently being heard in the Supreme Court. But work on the project has continued unabated.

In 1985 the Working Group met to finalise its report. The chairman Shri S.K. Roy expressed his opposition to the project. In 1986 two drafts of the final report were prepared by the experts. Both were rejected by the Chairman who asked for additional points to be incorporated.

The final report of the Working Group was submitted on 28 October 1986. It included a letter of dissent by the Chairman as the engineers from Roorkee University and the Chief Engineer of the Tehri dam project had, in their report, refused to support his stand that the dam was too risky. In September 1986 the Chief Engineer, Tehri dam project, the Chief engineer, Multi-purpose and hydroelectric projects, Irrigation Department, Uttar Pradesh and the Vice Chancellor of Roorkee University wrote letters to the Secretary, Department of Environment, requesting him to ignore the final report that was submitted by the Chairman.

The Department of Environment however refused to be pressurised by the UP engineers are recommended that the project was not advisable. When in October 1986 the final report of the Working Group was sent before a Committee of Secretaries, it was decided that the Central Water Commission should convene a meeting to give its opinion on the dam. Particularly on the issue of seismicity. The CWC convened a meeting of those experts who were known to be pro dam engineers. In spite of fears expressed by the Geological Survey of India, National Geophysical Research Institute and many independence experts, the CWC Committee resolved that (a) no major geological faults existed at the dam site; (b) adequate studies had been made to arrive at a safe and economical engineering structure to combat the risk of seismicity, and (c) adequate safety factors were incorporated in the design and construction of the dam and appurtenant structures. On 10 March 1987, the Minister of Environment and Forests observed in the Rajya Sabha that the Working Group had failed to make a categorically recommendation either approving or rejecting the project. Therefore, the decision about going ahead with the Tehri project was left to the Government.

Involuntary displacement of human population is always traumatic. Irrespective of the causes leading to such migrations the degree of suffering experienced by such people simply cannot be quantified in money values, and even in words it can be described only inadequately. In the case of natural calamities and wars one notices a sense of helplessness because the causes and consequences are so diverse and widespread that compensation and restoration become either intractable or apologertic.

However, this need not be the case when displacement is the consequence of preplanned development projects undertaken by the Government or other public authorities. But, unfortunately, ousting of people likely to be submerged under irrigation or hydel power dams is a classic case where hardships are imposed on people in spite of the 'pro-people' laws and policies proclaimed by the Government.

A liberal interpretation of the 1894 Land Acquisition Act itself indicate that the man objective must be:

(a) to ensure that in no case should the displaced person be economically worse off (say, measured in terms of his standard of living) as a consequence of rehabilitation or resettlement. preferably he should be more than equally compensated for his sacrifice towards a national cause;

(b) to ensure that, where possible, a village-community be resettled as an organic and integrated social entity instead of being scattered at random;

(c) to ensure that the displaced persons are integrated in the socio-economic hierarchy prevailing in the new locations rather than being forced to eke out a living in urban slums;

(d) to ensure the provision of community services and facilities like drinking water, wells, civic drainage, schools, medical dispensaries, road, places of worship, cremation grounds, "balwadis", playgrounds and community centres, etc.

The Uttar Pradesh 'Land Acquisition Act' has several other loopholes which are exploited by the project authorities. The value of land is calculated as on the date of Gazette Verification whereas the interests is paid only from the date of taking possession to the date of actual payment of compensation. Other errors of omission are due to the escalation of market values of land, undervaluation of land in Government records (because of the tendency to avoid 'court-fees' in case of land transaction). This results in unrealistically low rates of compensation. Therefore, the compensation paid to displaced persons is totally indequate to buy new land.

Our statesmen and legislators appear to be so obsessed with the welfare of Tibetan, Bangladeshi, Pakistani or Sri Lanka refugees into a "refugee-status" within their own country by unimaginative and often callous authorities executing developmental projects. Below is a critique of the Tehri Dam Rehabilitation Scheme based on the Detailed Project Reports and the Writ Petition filed by the Tehri Bandh Virodhak Sangharsha Samiti in the Supreme Court on 20 November 1986.

(1) *Land allotted as compensation :* The project authorities commenced the Scheme by allocating 2767 acre of land in the Dehra Dun area which was already reeling under severe pressure from tourism, limestone quarrying and urban expansion. Felling of trees from such a large area necessarily meant and aggravation of the problem of erosion and denudation of lands around Dehra Dun.

(2) *Rehabilitation of individual families as against villages as a whole :* In the urban areas, the families are more or less independent of one another. Commoditisation has reduced the dependency of one family on the other. In contrast in the villages, almost each family depends on the other. The social and moral obligations towards each other bind them into one cohesive whole.

The authorities are rehabilitating individual families and not the village as a whole. It is because rural society form such an integrated unit that it is so vulnerable to destruction. If one feature of the cultural pattern is changed, the other features change inevitably.

(3) *Divide and acquire :* The TDP authorities, faced with stiff opposition to their move to acquire land, have cleverly circumvented an ugly situation by offering lucrative compensation to the scheduled castes, the Dome and the Bajgis, who were to be displaced. As expected, they withdrew their opposition to the dam. The TDP authorities applied psychological

pressures too. For example, those who vacated the land late were threatened with not getting land as compensation.

Mission of ousting the people from the dam site completed, the Tehri Dam Control Board, vide order No. 885/ R.T.N.P/T-9/83 para 3 struck the final nail in the coffin on 30 June 1983. This order ruled that those families who could earn a livelihood without depending on land, be paid compensation in cash only, instead of land.

This is a shrewd way of excluding the oustees from any claims and Srisailam experiences are examples.

(4) *Monetary Compensation :* On paper, the principle of rehabilitation is : Land for Land, and Land for the Landless too. But while implementing the scheme, the principle is more like 'Cash for Land, and nothing for the Landless'.

Mere payment of cash is not rehabilitation. Moreover, the amount of cash paid as compensation is insufficient to buy land in other places because of the high rates. The demand generated by the displaced persons further increases the prices. The oustees being basically farmers lack the business acumen needed to set up a viable commercial alternative. Since they are not accustomed to having such large sums (relative to their usually small incomes) in a lumpsum, they are ignorant as to how they should spend it.

(5) *Faulty estimation of the number of affected people:* The project authorities estimated the total affected population in 1981 as 46,000. Using the Census Office figures, the total number affeccted for 1981 is actually 70,000.The discrepancy in figures for Tehri villages is the result of the TDP authorities not accounting for the villages situated on and around the rim of the reservoir to be constructed.

(6) *Lack of Public Relations* : The majority of populace to be displaced consists of adivasis, tribals, scheduled castes who have a unique lifestyle. Their villages do not form compact household clusters but are scattered all over the region.

(7) *Monuments and Archaeological buildings:* In the measures for rehabilitation of rural and urban people,the project authorities mention: 'Rehabilitating all the religious, historical and archaeological buildings and monuments in the new town. It is difficult to visualise how the dam authorities propose to rehabilitate monuments and buildings. A monument or a religious place of workship acquires importance because of many factors like its location, historical background, beauty and/or architecture, or, because it has a special significance.

(8) *Compensation for houses* : The present procedure evaluates the "current worth" or "value after depreciation" for determinating the amount of compensation. This concept is faulty and also contrary to the spirit of the Land Acquisition Act 1894 which rules that the displaced persons, whose land and property are acquired involuntarily for a public purpose should be compensated in such a way that in his rehabilitated condition he should be at least as well off as before.

4 World Population Growth

One of the major areas of interest and concern for ecology is the scientific study of population. This includes the analysis of population structure and composition, population growth patterns, the natural control of population, and the interaction between population of man, animals and other organisms. Population studies are fundamental to many human problems in agriculture, pest control, wildlife management, and economic planning. The purpose of population research is to understand the interplay of factors influencing population range, and ultimately to predict and regulate the course of populations. Man obviously wants to limit or reduce populations of organisms, whereas he has wish to maximize within reasonable limits valuable and productive animals for aesthetic or economic reasons.

The world population was almost stationary during Stone Age, an estimated 10 million people. Sometime between 8000 B.C. and 6000 B.C. man learnt the art of growing food, and so could support a large population. As a result the human population increased 50 times in 7000 to 9000 years. In the next two centuries the population doubled and reached an amazingly high figure of one billion. It took just 100 years to double the population to reach four billion in 1975. Alarmingly, in 12 years it reached the five-billion mark on July 11, 1987. At the present rate of growth we will add another billion in the next 12 years, unless drastic measures are taken. The present world population of 5.43 billion, which is growing at the rate of 97 million people per year will touch 8.5 billion by the year 2025. About 95% of the population growth will be in the developing countries. At present we add about 500 million people in about six years times, a figure which took thousand of years to reach before 1950 AD. About 3,42,000 babies are born every day in the world and about 1,5,000 persons die, leaving a net increase of 2,07,000. Thus within a week's time there will be nearly 1.5 million more people. The UN population Fund in a recent report says that in the next 10years there will be one billion more people on Earth, making a total of 6.4 billion.

There is no doubt in the fact that human population are drastically out of balance. For thousands of years prior to the seventeenth century, the population of the world was less than 500 million people. Human population were limited by high mortality and even in the great civilizations of Egypt, Greece and Rome, expectation of life probably did not exceed 30 years. The class checks on human population were disease, famine and war. In the fourteenth century, at least 25 percent of the adult population of Europe died in epidemics of bubonic plague and between the years 1348 and 1379 A.D. England's population was reduced almost 50 percent, Cholera, typhus, small pox, malaria yellow fever and sleeping sickness have all been among the great killers of man throughout history. Several of these diseases, especially malaria and yellow fever, severely limited the abundance and distribution of people in tropical countries until the advent of modern medicine and public health. Asia has traditionally been the great stage for famine though starvation; death also occurred in Europe.

The famous Irish famine of 1850 resulted in the death of more than 12 percent of the adult population between 1846 and 1851. In China, in 2000 years preceding the 20th Century, over 1800 separate famines were recorded, many of them killing million of individuals. More recently, 4 million famine deaths occurred in China in 1920-21 more than 5 million is Russia from 1918 to 1923; and over 2 million in India in 1943. Almost agriculture and transportation have alleviated famine mortality, malnutrition continue to be significant demographic for, particularly since it accentuates illness and mortality from infectious disease. War has also been a major mortality factor throughout history,not only by causing direct battle casualties, but also by stimulating disease and famine. Great epidemics and massive crop failures, have often been triggered by the disruptive condition of warfare. The casualties of war have frequently involved as many civilians as soldiers. Some historians believe that the Thirty year war in Europe between 1618 and 1648 resulted in the death of 30 percent of the inhabitants of Germany and Bohemia. Despite the combined action of disease, famine and warfare, several major development in the sixteenth, seventeenth and eighteenth centuries stimulated population growth and eased the heavy mortality burdens of man. The opening of the New World provided a vast new area of space and nature resources of economic exploitation. The industrial revolution greatly increased the productivity and control of infectious diseases. Between the birth of Christ and the seventeenth century, the world's population took approximately 1600 years to double. The next doubling as achieved within 200 years, and the next from 1 to 6 million people within 80 years, from 20 to 4 million people were completed by 1975 a span of 45 years and at the current rates of growth the world population will double again within another 35 years.

The following table indicates the populations growth rates:

Table 1

Region	Population (billions)			%Annual Growth rate		
	1985	2000	2025	1950-85	1985-2000	2000-2025
WORLD	4.8	6.1	8.2	1.9	1.6	1.2
Africa	0.56	0.87	1.62	2.6	3.1	2.5
Asia	2.85	3.55	4.54	2.1	1.6	1.0
Latin America	0.41	0.55	0.78	2.6	2.8	1.4
North America	0.26	0.30	0.35	1.3	0.8	0.6
Europe	0.49	0.51	0.52	0.7	0.3	0.1
U.S.S.R.	0.28	0.31	0.37	1.3	0.8	0.6
Oceania	0.02	0.03	0.04	1.9	1.4	0.9

Currently the world's population is increasing at approximately 2 percent per year—a net increase of about 1,300,000 per week. Every day of every week, the world's population increases by more the 207,000 people. All this is primarily a product of man's increased life span a reduced mortality through modern medicine and agriculture. Man's fertility patterns were basically established during periods of much higher mortality rates thousands of years ago and they have remained high in the intervening year.

The developing countries least able to afford rapid population growths are experiencing the greatest growth. The population is not evenly distributed around the world. More than half the world's population regions of the total world, i.e. Europe, North America, former USSR and Oceania had only about 22% of the total world population in mid 1988. The remainder was distributed in the developing region of the world. The growth rate is most rapid in Africa. In the coming decade, the population there will rise from 640 million to 950 million and it will be the highest regional growth rates the world has ever known. According to projection, countries like Ethiopia, Nigeria, Zaora and Kenya will double their populations within the next 20 years. In North Africa the growth rates are slower, but th figures are still rising fast. South Asia and Africa together will account for over half of the increase. Southern Asia's population will rise from 1.2 to 1.6 million. In Eastern Asia the population growth rates have declined rapidly in the past decade, but the number will still increase from 1.3 to 1.5 million people. Several Arab countries have very high population growth rates and the 12 countries in West Asia will increase in their population to 131 million by 2001. The population; of Latin America and the Caribean is projected to increase by 100 million from 440 to 540 million. In Central and South America, most countries have annual growth rates in excess of 2%.

In addition to the large concentration of population in developing countries, the percentage of population under 15 years and over 64 years of age what is called the dependency ratio is also very high. The population under 15 year is about 45% in Africa, 35% in Asia and 39% in latin America, whereas it is only about 23% in North America 22% in Europe and 29% in Oceania. However, the percentage of population over 64 year is higher in the developed regions of the world. North America has about 11% of its population over 64 years. Europe has about 13% and Oceania about 8%. It is only 3% in Africa and 4% in Asia and latin America. The high dependency ratio in the developing countries means that age group between 15-64 has to support a large population with requirements for housing, clothing, education, health and other social service. The large young population in developing countries also has demographic implications in the near future.

Asia and Africa, and the regions which will account for about 75% of the world's population in the next 35 years or so, and will have the greatest proportion of poor people. In South Asia 51% of the population lives below the poverty line and in sub Saharan Africa 45% Malnutrition, ill-health, illiteracy and the lowest life expectancy, accompanied by inadequate technology and economic poverty, are predominant in South Asia and Africa. It is not very difficult to imagine the miserable plight of the people in their regions in the next century.

The UNEP has warned that continuing high population growth will result in continuing poverty and increasing environment damage and will lead to higher levels of urban growth and international migration. The population report support this that it will be impossible to keep pace with population trends. If the population growth continues as expected, it says within 10 years there will be more than 200 million children who will receive no education. What does this mean in terms of human welfare and environment quality?

It usually means that many countries have excessive demands for products and services which cannot be met. It is exceedingly difficult, if not impossible, for those countries with the highest rates of population growth to meet the requirements of their people for food, housing, jobs and the physical amenities of life, and it is virtually impossible for them to maintain adequate services in education, medicine, public health and social welfare. Thus we emerge with such problems: (1) 60 percent of the world's people are still inadequately fed, receiving a percapita calorie, intake less than that recommended by the UN (2) 80 percent of the children of India shows evidence of malnutrition; (3) there are only 9 acres of land surface available for every person on the earth, and only 26 percent of this is available for agricultural production; thus only 1.2 acres of potential cropland are currently available for each person and this is decreasing each year (UN States of Hunger Report); (4) many countries, especially in Asia and Africa, still have an illiteracy rate exceeding 70 percent; (5) in many developing nations, there is only one physician for every 5,00 to 10,00 people, a ratio which makes adequate medical care impossible. This list could be extended considerably to indicate that many parts of the world are currently living under severe population pressure. While the western nation gain strikingly in prosperity, most of the world's people face increasing poverty and population pressure.

When to global picture is considered, the outstanding fact exists that we can not provide a quality life or maintain a quality environment for the majority of the world's people at the present time, let alone provide for more than 1,600,000 new people every week. Ecologically, this represents a highly unstable condition, it cannot persist indefinitely without major corrective or compensating changes. These grim facts coupled with increasing global communication, rapid social and technological change and rising expectations throughout all nations add fuel to the fire of social and political instability. As noted by Eric Hoffer in The Ordeal of Change (1963):

> "...... a population subject to drastic change is a population of misfits-unbalanced, explosive, and hungry for action. In other words, drastic change under certain conditions, creates a proclivity of fanatical attitudes, united action, and spectacular manifestation of fluting and defiance. It creates on atmosphere of revolution.

The next decade may decide the future of the earth as hesitation for humans. Already our impact has been sufficient to degrade the soil of millions of hectares to thin the ozone layer threatening the rain-forests and to initiate global warming the full consequences of which cannot yet be calculated. At some point in the not too distant future the changes in the environment may cross the threshold of catastrophe. Again we see the intimate relationships between ecology and human affairs. It does not imply that ecology can provide the answers to all these problems but surely a failure to recognize their ecology background invites certain disaster.

India's population, as on March 1, 1992 stood at 843.93 million(437.60 males and 406.33 million females). The second largest population country, India is now the home of 16 percent of world's population. With this, every sixth person in the world is now an Indian. The country however accounts for 2.42 percent of the total world area. Population counts in India in the modern sense started in 1881 since then, a population census has been conducted regularly every ten years. A census of population was conducted during 1871-72 but it was nosynchrononus. Moreover, the 1972 census omitted served territories, the population of which totaled some 33 million in 1881.

The population of India as recorded at each decade census form 1901 has grown steadily except for a decreased during the decade 1911-21. In absolute terms the country population has increased by 160.60 million from 680.3 million in 1981 to 843.9 million in 1991. This is ten times the population of Australia and is more than the population of Japan. An encouraging feature revealed by the census result is the decline in the growth rate of the population in the last decade compared to 1971 to 1981. It marginally decreased from 24.66 percent recorded in 1971-81 to 23.50 per cent during 1981-91. In most state the growth rate declined during the decade. However, Andhra Pradesh, Arunachal Pradesh, Madhya Pradesh, Maharashtra, Nagaland, Jaipur, West Bengal, Daman and Diu, Lakshadweep and Pondicherry which account for one third of the country's population recorded increase in growth rate. Nagaland registered the highest growth rate of the 56.86 percent while Kerala registered the lowest 13.98 percent. Uttar Pradesh continues to be largest state population wise with 16.44 percent people of the country living there (138,760,417), followed by Bihar with 10.32 percent (86,338,853) and Maharashtra with 9.33 percent (78,706,719).

West Bengal occupied fourth place accounting for 80.06 percent (67,982,732) of the country's population followed by Andhra Pradesh with 7.86 percent (66,304,854) Bombay continued to be most populated city of the country with urban population of 12.57 million, Calcutta taking the record place having 10.86 million, Delhi ranks third with a population (urban) of 8.38 million followed by Madras 5.36 million.

One of the important indices of Population concentration is the density of population. It is defined as the number of person per square kilometer. The population density has gone up from 216 in 1891, to 267 person in 1991. In 1901 it was 77 person. Density is the highest in Delhi (6.319) followed by Chandigarh (5,620). Lakshadweep which has the smallest population size is the third most densely populated area with a density of 1,615 person per square kilometer , Pondicherry ranks fourth (1605) followed by Daman and Diu (906) Arunachal is the most sparsely populated area with a density of 1,615 person per sq kilometer.

The ten heavily populated districts of the country are Calcutta Madras, Greater Bombay, Hyderabad, Delhi, Chandigarh, Maoe, Haora, Kanpur City and Banglore. All of them have density of above 2,00 person per sq km and 5.0 percent of the country's population lives in these districts. The average density of these 10 districts. The average density of these 10 districts is 5,791.

In any study of the population, analysis of the sex composition plays a vital role. The sex composition of the population is affected by the differentials in the mentality conditions of males and females, sex relative migration and the sex ratio at birth.

Table 2
Sex Ratio (Females per 1,000 Males) in India, 1901-1991

Census year	*Sex ratio*
1901	972
1911	964
1921	955
1931	950
1941	945
1951	946
1961	941
1971	930
1981	934
1991	929

The sex ratio is defined as the number of females per 1000 males. In India it has generally been adverse to women. The ratio has also declined over the year except in 1981 when it slightly improved to 934 from 930. In 1991, there has been a fall by five points to 929 per 1000 males. But Kerala represented a different spectrum. The State has a higher number of females than males, 1040 females for 1000 males.

Many scholars have inferred from the actual experience of several European countries and the United States of America that populations historically pass through different more or less well-defined stages in terms of their levels of fertility and mortality. As a generalized explanation of process of mortality and fertility decline in those countries, the transition theory refers to a stage of high fertility in those countries, the transition theory refers to a stage of high fertility and high mortality in simple, mostly agarian, societies which finally transforms itself to a condition of low fertility and low mortality after passing through several other stages.

Today the transition theory it sought to be equally applicable to the less developed counties which are still in the early stages of demographic change. In effect this theory has come to be increasingly regarded as theory that might anticipate the future demographic trends in countries which are currently having high population growth due to rapidly decreasing death rate and high but more or less stable or only slowly declining birth rate. Although there are several expositions of demographic transition theory, but the one dividing the total period into five stage is more commonly used. These stages are as below;

(a) The high stationary stage characterised by high birth rates and high death rates;

(b) The early expanding stage with high birth rates but declining death rates;

(c) The late expanding stage with falling death rates but more rapidly falling birth rates;

(d) The declining stage with low mortality now natality and deaths exceeding births.

The first stage which is characterised by agrarian low income economy is marked by relatively stable birth rate at the maximum fertility level found in societies not using contracepitives, and the death rate is found to fluctuate in response to conditions of prosperity or famines, epidemics and other disasters. Then, as the economy progresses to become more interdependent specialised and market dominated with the generation of agricultural surplus, the average death rate begins to decline under the impact of better organisation and improved medical knowledge and care; however, the birth rate continues to remain almost stable at the earlier level. This leads to an increasing growth rate of population which is the early expanding stage.

As the death rate continues to decline with further improvements in living conditions birth rate also starts declining first slowly and somewhat later at a faster speed making the 'late expanding stage'. In this stage the growth rate starts getting smaller and smaller than what it was in the early expanding stage. In the fourth stage when further reduction in death rate becomes harder to attain, the death rate stabilizes at a low level(between 8-12 per 1000 persons), the birth rate as approaches equality with the death rate and a more gradual growth rate is established when both birth and death rates are low but birth rate fluctuates some what from year to year in response to voluntary decisions of the people.

In the last stage of demographic transition which started to appear in recent years in a few European countries like Germany, Sweden, Switzerland etc., the birth rates has declined even below the replacement level and there may not be enough people in the next generation to replace the present one.

For India, we now have a history of around 100 years of changes in birth and death rates.

The first stage of demographic transition in India continued till about 1920 when both death rates and birth rates were high. This stage was also marked by purely agrarian economy with hardly any significant industrialisation. High mortality levels were hardly and for the variable growth rates of the population during the 30 or more years before 1921.

A severe famine affected large areas of the country in 1896 and 1897. In the Bombay Presidency (excluding the princely states) especially, the effects of the famine were aggravated by a severe plague. As a result of the two scourges, the 1901 census recorded a population loss of about 2 percent from the 1891 level of 18.8 million.

Table 3
Birth and Death Rates in India from 1891-1901 to 1989

Year	Birth rate	Death rate
1891-1901	48	48
1901-1911	49	49
1911-1921	49	49
1911-1931	47	37
1931-1941	45	33
1941-1951	43	31
1951-1961	44	26
1961-1971	42	20
1971-1981	37	15
1980	33.7	12.6
1981	33.9	12.5
1982	33.8	11.9
1983	33.7	11.9
1984	33.9	12.6
1985	32.9	11.8
1986	32.6	11.1
1987	32.2	10.9
1988	31.5	11.0
1989	30.5	10.2

Like the 1891-1901 decade the next decade also witnessed several local famines and a severe one in 1907 in most parts of Uttar Pradesh. Plague was in evidence in the Bengal and Bombay Presidencies, and both plague and malaria were widespread in the Punjab and Uttar Pradesh where population growth was negligible. Yet, because the country as a whole suffered from famines less widely and for shorter durations, there was an appreciable rise in the population growth rate compared to the previous decade. During the 1911-20 decade India suffered from an influenza epidemic that caused an estimated 7 percent of the total population to die and led to an absolute decline in the population of the Indian Republic (according to the present boundaries) by 0.77 million at the 1921 census.

The second phase of demographic transition in the country started with the beginning of the 1920s and has spread well up to 1971. This is also the period when industrialization began in the country and took some roots. During this period the major causes of high mortality-famines and epidemics-have been gradually brought under control and between 1921 and 1951, the country witnessed a gradual rise in population growth rates. The decline in death rate became sharper during the 1950s without any appreciable decline in birth rate resulting in a population growth rate of 1.96 percent per annum during 1951-61 in comparison to 1.25 percent during 1941-51 decade. During1961-71 the birth rate still remained quite high and due to further decline in death rate the growth rate increased to 2.05 percent.India seems to have entered the third stage of demographic transition since 1971 when birth rates has also started declining. During the birth rate leading to a plateau in the population growth rate during the 1960s and the 1970s. In the 1980s, probably the decline in birth rates was slightly greater than that in mortality; consequently the growth rate during this period is slightly lower than in the previous decade. It is now hoped that there would be faster declines in birth rate than in death rate in the next couple of decades. It is also hoped that India will be at the point by about 2020 if not earlier.

Only one in ten people lived in cities when the century began more than half will by the century's end. Today the vast bulk of urban population growth occurs in developing countries. The populations of Third World cities, now doubling every 10-15 years, overwhelm government's attempts to provide clean water, sewage, adequate transport and other basic services.

Cities in Industrial countries are gaining people much more slowly, yet even as their growth rates slow or core population decline, these urban areas continue to spread out ward. Industrial and developing countries alike, chaotic, uncontrolled urban growth-whether measured is risking numbers or in the amount of space humans spread out upon—draws on ever more land, water and energy from surrounding regions to meet people's needs.

The way cities physically evolve—and the way their development is planned has profound impact on human either recognize the limits of the natural environment or it can destroy the resources on which current and future societies depend, it can meet people's needs equitably or it can enrich some while impoverishing or endangering others. The world needs an urban planning ethic that is sensitive to these environmental and human dimensions.

According to 1981 census, India's urban population was 160 million ranking fourth among the countries in the world in terms of absolute size of the urban population, the first three being China, U.S.S.R. and U.S.A. Though the level of urbanization measured as the percentage of urban to total population has increased from 11 percent in 1901 to only 23 percent in 1981 the number of people living in urban areas and multiplied six folds during the period from 26 million, constituting more than one third of the country's population of 926 million at that time.

The growing chaos in urban areas are due to rapid immigration rather than natural increase. A total of 15.7 million persons have migrated from rural to urban areas in the country during the decade 1971-81. The immigrants in the reverse direction i.e. urban to rural numbered. 6.4 million giving rise to a net addition of 9.3 million people in urban areas. There are 218 class I cities in India with more than a lakh population each and these account for about 60 per cent of urban population. Among these, there are 12 metropolitan cities viz., Calcutta (9.16m), Greater Bombay (8.22m), Delhi 5.7m), Madras (4.27), Bangalore (2.9m), Kanpur (1.68m), Pune (1.68m), Nagpur (1.29m), Lucknow (1.0m) and Jaipur (1.0m). With the data on class I cities and the emergence of new towns on the eve of the 1991 census, we can examine the likely level of urbanization in India. It is well known that the share of Class I urban population has constantly agglomerations and cities in the total urban population has constantly increased over time.

If one looks at the total urban population has constantly increased over time. If one looks at the trend in India (excluding Assam) from 1951 to 1981, this share increased from 44.7 per cent to 60.8 per cent, the increase being 6.8 per cent during the 1970s. The increase in the share was higher during the 1950s since a very large a number of small towns (857 in all) were declassified on the eve of the 1961 census due to introduction of a restrictive definition of urban areas. In contrast, there was a net addition of 895 towns (or 782 urban agglomerations and towns) between 1971 and 1981 which involved reclassification of almost 250 towns declassified on the eve of the 1961 census. Addition of a large number of small and medium towns during the 1980s brought down the pace of increase in the share of population of Class I cities in 1981.

As the number of new towns indentified on the eve of the 1991 census is quite large, one may expect a substantial increase in the population of medium and small towns in 1991. If, however, a large number of them appear as part of urban agglomerations of Class I cities, their population will be counted as population of Class I cities.

An examination of the list of new towns and of cities and urban agglomerations indicate that a large number of new urban settlements have appeared as part of urban agglomerations. Accordingly, we may expect the share of the population at 65 per cent. This is in view of the fact that 80 Class I cities have been added during the 1980s. We may also not forget that the decadal growth rate of Class I cities slightly declined over the decade compared to the previous decade.

Assuming that the proportion of population of Class I cities in the total urban population of 1991 is 65 per cent, we get the total urban population as 213 million excluding that the Assam and Jammu and Kashmir. If we add the 1991 population of Class I cities of Assam, the estimate of total urban population would remain at the 1981 level (that is, 60.5 per cent), the 1991 urban population (excluding Jammu & Kashmir) works out to 230 million, almost the same as projected by the Expert Committee on Population projections in 1984.

Thus, the urbanization rate in 1991 varies between 25.6 per cent to 27.5 per cent. Keeping in view that the number of metropolises has increased from 12 in 1981 to 23 in 1991, and that the number of Class I cities has increased from 216 to 296 during the same period, one may expect a higher share of Class I cities in the total urban population than 65 per cent. If this is found to hold good when the data on rural and urban population are released, the total urban population would be less than 214 million. Population in the twentieth century has exploded particularly in the developing countries. Infact, before 1800, no country was predominantly urban. But by the year 2000, over 50 per cent of the world's population will probably live in urban places and there will probably be more than 250 cities of over one million inhabitants. There has been a world wide trend towards urbanization, the process characterized by the movement of people from rural to urban settlements, from small towns to large cities, and from large cities to the suburbs. This immigration is caused by and in turn leads to profound societies change. Rural areas are depopulated, while urban areas towns, cities and urban zones—become more densely populated. Since most of the cities are growing in an unplanned way there arises a number of problems such as housing, traffic, shortage of open space & pollution.

The Indian cities are growing both vertically (multi-storey building culture) and horizontally (invasion of the peripheries). In spite of cities expansion there is growing shortage of house (3.8m), growth of slums and squatter settlements, creating interpersonal and intercity imbalances, breaking down services and acceutuation of the problem of urban renewal. Slums are now becoming permanent features and are proliferating and sheltering larger population every year. They are having small tenements, narrow lanes, lack of sanitation, water supply, electricity, unhygienic environment, etc.

Due to realization of the futility of spending money on their clearance, the Government had decided to accord them recognition and gave their residents whatever amenities the funds allowed. Bombay is estimated to have the largest slum population of 3.3 m followed by Calcutta 3.2m, Delhi 2.6m, and Madras 1.8m. With an unabated inflow of job seekers, these cities are expected to see continuing rise in their slum population. By 1990 when Calcutta will have a population of 12.5m and its slum will be harboring 4.4m Madras population will be 11.8m, 9.7m and 6.0m and of them 4.1, 3.1 and 2.1m slum dwellers. Amongst States, Uttar Pradesh will have the largest slum population of 6.5m followed by Maharashtra 6.2m, West Bengal 4.9m, Tamil Nadu 4.2m, by the end of the decade. The Government's initial strategy was to clear the slums and rehabilitate their dwellers. This had to be given up by the end of the sixth plan as there was constant shortage of funds and developed land. The emphasis was shifted to an on going central scheme known as "Environmental Improvement of Urban Slums," envisaging a package of services for the immediate removal of unhygienic environment of the slums. The services included drinking water, community bathrooms, street-lighting, sewerage, storm water drains and street paving.

The traffic problem in urban areas is one of the greatest problems of the day. Most of the cities and towns are growing in unplanned way. Due to lack of proper enforcement agencies encroachment in all parts of the cities is going rapidly and due to political interference, authorities become in competement. Most of the roads are narrow, haphazard and ill-maintained. Traffic problem is getting bad to worse because of high density of population desire to have personal means of transport and to live near the city, mixed traffic, lack of civic sense, ill-maintained road, inadequate expansion in the existing road network poor regulation of traffic, etc.

Delhi had 12,000 vehicles in 1951 and the number had shot upto 9 lakh in 1986 and this is expected to reach 40 lakh by 2001. Apart from the fast moving vehicular traffic, the problem on the city roads is being further aggravated because of 50,000 or more slow moving vehicles such as tongas, rickshaws and bullock carts and 20 lakh cycles. There is also a growing trend of road accidents. There were 6,037 reported road accidents of which 1,162 persons lost their lives in 1986. According to National Transportation Planning and Research Centre, Trivandrum in every four minutes there is one road accident in India, killing or injuring one person on an average.

Due to increasing commercial value of the land in cities, open spaces are diminishing, even though green spaces are the lungs of the cities, the guarantors of a healthy and aesthetic environment. In Japan, urban parks are classified under two main categories; parks with structures or city parks and zoned green spaces, which are further subdivided into green space conservation zones, agricultural green space and special historical land space zones/scenic beauty conservation zones. City park land includes playgrounds, open spaces, planting strips along side road and other lands awaiting full development. It is commonly said that these parks become the pursuit of undesirable activities including sex, crime, gambling, etc. In spite of these negative aspects, it provides thousand times much needed benefits for the city dwellers.

There is a concept to have at least 20 per cent area of the cities under open space in the form of green belts, parks, playgrounds, etc. But it has become an utopian concept. In 1973, master plan of Allahabad was approved and much of the areas were put under green belts. But due to illegal spraw of colonies most of that very areas were converted into concrete belts. The green belts concepts are also not getting proper place in the city environment improvement scheme because of lack of financial support, public interest and professionally qualified staff. Besides this, there is not definite framework for parks and recreational areas in relation to national goals.

City parks and green belts are developed mostly by Municipalities or Corporations, Social Forestry and Horticulture departments. Private organizations, voluntary organizations, industries should also come forward to set city parks and green belts. Vacant lands acquired by the Corporation are to be sued not only for commercial or residential settlement but also for establishing big parks. Local communities must come forward to participate in the management of parks and can also raise funds for it.

Country parks can be established in rural areas to provide a better place for recreation and enjoyment of rural people. In addition to parks and gardens, landscaping measures should be applied on flyovers, overhead bridges, wall and other structures, such as the growing of creepers, climbers, scrubs and transplanted trees. There is a need for making legislation for City and Country Parks Act for the forceful implementation of the concept of green belt.

Virtually all cities share some land related concerns, such as congestion and pollution from motor vehicles, lack of affordable housing, and the cancerous growth of blighted district. Of course, prescriptions for better urban planning are not the same for all parts of the world. Many land use issues in industrial countries seem mainly concerned with the quality of urban life, and those in Third World cities are often questions of life and death. But all cities, whether surrounded by affluent suburbs or makeshift shanty towns, now need to plan land use far more carefully than in the past before the developing world's urban crises than into catastrophes, and the industrial world' urban crises turn into catastrophes, and the industrial world's problems become issues of survival. The extent to which cities control the use of land varies widely. Japan and Western Europe have the world's most comprehensive urban land use controls; their cities are generally compact, laid out for efficiency in transport, water and energy use, and building materials. Japan protects farmland with strict zoning and tax policies. Its suburbs are of moderate, not low, density. Much of Japan's urban activity is concentrated in Tokyo and along the dense Tokadio corridor between that city and Osaka. This compact, linear form has allowed great efficiency in transport, including development of the world's first high-speed train, which has been running since 1964.

However, the country's 33 million automobile—nearly twice as many as in all other countries combined—still plague Japanese cities with extreme traffic congestion, especially in the Tokyo area. Western Europe has a long tradition of activity controlling land use so that the small amount of available space serves the public's interest more than that the private developers. Paris, a city for nearly 2,000 years, has planned and regulated land use since the Middle Ages. Many of England's urban areas, still fulfilling a far sighted decree by Queen Elizabeth I in 1580, are ringed by green belts intended to protect farmland and prevent sprawl. But in post war decades the automobile has had a profound effect on Europe's compact character, fuelling suburbanization.

In the fifties and sixties, some older European cities embraced the U.S. model, demolishing corner shops and cafes to make way for urban freeways. Later, planners tried to contain suburbanization by diverting growth into smaller satellite towns. Since the mid-seventies, they have turned their attention to revitalizing shrinking inner cities. Control of land use in Eastern Europe and the Soviet Union has not been as effective as might be assumed for centrally planned economies.

Urban planning therefore is not well coordinated with economic decisions, and sometimes reflects rivalries among government ministries. Extreme water air, and soil contamination from poor placement and regulation of polluting industries is a critical concern. Although urban planners in these countries historically have supported public transport and promoted urban densities high enough to make it viable, in recent years and use has become less efficient.

Nearly all new housing has sprung up at the edges of cities, with few accompanying shops, jobs, or schools, resulting in excessively long commutes. On going political reform and the region's cities will become more automobile-oriented, perpetuating the environmentally destructive patterns of western industrial cities. Among industrial regions, North America and Australia have the weakest planning traditions. Governments on these continents have done relatively little to guide development beyond separating industrial areas from those zoned for commerce and housing. Many have enforced low-density zoning in an effort to curb urban expansion—a move that, ironically, achieves just the opposite.

Zoning codes that restrict residential density, usually by requiring each house to occupy its own large lot, have forced development to consume even greater tracts of open space. And to accommodate new, outlying communities, roads have been extended further outward, attracting even more spread out growth. Even older U.S. cities are becoming more dispersed now than in the pre-automobile area. Growth in the New York metropolitan region has turned to sprawl; while population has grown only 5 per cent in the past 25 years, the developed areas has increased by 61 per cent—consuming nearly quarter of the region's open space, forest, and farm land.

Many people move out to the suburbs and exurbs seeking open space and bounds with nature that come only in a rural setting. Yet most of these residents continue to maintain urban lifestyle—communiting to jobs in the city and demanding urban amenities from a suburban shopping mall near their home. The result is neither urban nor rural living but a destructive compromise that the environment sustain.

Developing countries have the loosest controls over how cities develop. Local governments in Asia, Africa, and Latin America often have neither the authority to guide land use nor the funds to provide basic services. Municipalities are fiscally dependent on central government, which are themselves struggling under crushing debt burdens.

With few exceptions, urban planning is a relatively recent phenomenon in developing countries, even in the capitals. In India it reportedly began with Delhi's Master Plan in 1962. In Nigeria, the first systematic effort to plan cities did not begin until 1971; the capital, Lagos, still has no designated metropolitan government and no formally defined urban area. The Lagos State Ministries that have their own priorities.

The Third World is burdened by several enormous, rapidly growing cities—including-Sao Paulo, Shanghai, and Mexico City—whose sheer size and instability create problems on an entirely different scale. As early as 1964, researcher Ronald Wraith used a single word to denote these giant cities racked with pollution and rimmed by shanty-towns: "Megalopolis-the city running riot with no one able to control it." Since then, megacities have become increasingly characteristic of the developing world. In 1950, only three of the world's 10 largest cities were in the Third World by 1980, seven of them were there. Although skilled land use planning is badly needed in these megacities, its effectiveness is limited in the absence of other kinds of change. One reason is that most of the physical growth takes place in illegal unplanned squatter settlements, rendering useless even existing mechanisms for guiding land use. Called fevelas in Brazil bidonvilles in Freńch-speaking West Africa, Spanish in the Middle East, and kampungs in Indonesia, these illegal communities hold 30-60 per cent of the population of many Third World cities.

The unhealthy conditions in these settlements can only be addressed fully through extensive economic and social reforms that attack the root causes of poverty not just in cities but also in the rural areas that urban migrants abandon in search of economic opportunity. The effectiveness of land use planning is also limited in the absence of family planning programs. (Only in Africa does migration account for the increase in much of Asia and Latin America results from the fertility of people already in cities).

Finally, planners in the Third World's giant cities faces colossal environmental problems, from deadly air pollution in Mexico City, to indiscriminate dumping of toxic wastes in Alexandria, to the actual sinking of cities such as Bangkok, Jakarta, and Shanghai due to overdrawing of ground water.

Crowding is a conspicuous ecologic force impinging upon modern man. It results from more than a simple increase in population density. It is also a product of activity patterns psychological needs, and social contact. Crowding implies a force, a pressure and psychological reaction. No clear definition of crowding is possible at the present time; it remains an operational concept with many subjective qualities. Crowded population are, however, easier to recognize than to define, and they are studied more carefully, better definitions may become possible. For the time being, we might define crowded populations as those in which there seem to be excessive number of individuals per unit space in relation of the activity of the industrials and the quality of the environment.

With an operational concept of this type, ecological studies have shown that some rodent populations may be "crowded" in relation to their social behavior and habitual requirement at a density of 5 individuals per acre, where as other at the same species may not be "crowded" until they have a density of several hundred per acre. Species differ in this regard, and so also individuals and social groups within the same species.

A western cattle rancher may feel "crowded" at a density of 10 homesteads per square mile whereas a modern suburbanite may feel relatively "uncrowded" at a density of several hundred homes per square mile. The western rancher ranges several square miles per day over what he considers his personal property and he does not welcome potential competitors in his territory.

Conversely, suburbanite restricts his daily movements, apart from going to work and to shop, to a lawn around his home, and he does not view neighbors as a competitive threat. Hence, in both animal ecology and human sociology, it is important to distinguish between population density per se, and crowding, since the latter depends so much on behavioral factors. The spatial requirements of man are obviously very variable, and result from both individual difference and social conditioning. It is impossible to state what the optimal spatial and social needs of man are as those of the Japanese city dweller ar much different than those of the Kansas farmer.

Nevertheless, it is safe to assert that man is now subjecting himself to increased crowding through the combined forces of population growth, urbanization, transportation and mass communication. Man's perception of crowding has skyrocketed upwards in recent years.

The growth of urban traffic and the increased number of people in public buildings, schools, hospitals, stores, exhibitions, recreational areas, etc., have all contributed to the feeling of population pressure. Through the mass media or audios, television and newspapers, individuals are vastly more aware of worldwide events. Social pressures at any point of the globe are quickly felt by all nations.

Local crime, violence, and warfare now become worldwide worries in a matter of hours. The people of one continent feel threatened by those of another continent thousands of miles away. The inevitable result has been a great magnification of social anxiety and population pressure.

The world has thus become more limited in its psychological space, and man more crowded in a social and behavioural sense. How does one evaluate crowding as an ecologic force? It definitely requires knowledge of at least three major parameters: (1) population density, (2) environmental structure, and (3) social behaviour in relation to each of the possible in both animals and man, but in both, especially man, it is often possible to isolate crowding as the key variable. Many other factor are interwoven.

Crowded human populations are often so characterized by poverty, malnutrition, lack of educational and recreational opportunities, varies environmental hazards and unstable social patterns that it is virtually impossible to separate crowding itself as the one key variable. The essential question becomes, how does one measure and analyze crowding per se apart from all it related phenomena? It may also be important to distinguish between crowding of various origins.

Since man often seeks social contact and stimulation, does not crowding contain constructive and creative social elements, or is it inevitably a factor in increased social problems and environmental deterioration? From man himself the answer to these questions are confusing and uncertain. The behavioral sciences of man have just begun to investigate human spatial behavior.

More direct clues have come from animal studies where experimental work has been possible. The simpler social systems of many animals have permitted more direct observation of behavior, more ready identification of variable, and more penetrating analyses of causes and effect relationships. In animal studies it has been possible to distinguish between extrinsic crowding; that crowding imposed on the animals as an external force, as in captivity, compared to intrinsic crowding in which dense aggregations result from the social behavior of the animals.

A similar distinction is sometimes possible in human populations, but frequently it is difficult to discern if people are crowded by choice or the external pressures of economics and society. It was demonstrated more than 30 years ago by Professor W.C. Alee and his students of the University of Chicago that most species of animal have optimal levels of crowding, above which and below which deleterious effects occur. These studies showed that certain levels of crowding are necessary to maintain normal biologic function and social activity.

Goldfish, for example, grow most rapidly and live longest in social groups of certain densities compared to isolated fish to fish in smaller groups. Fresh water shrimp survive longest and tolerate harmful environmental conditions most successfully at certain optimal stage of crowding. Similarly bobwhite quail survive harsh winters best and reproduce most successfully in the spring in conveys of certain optimal size.

Beyond optimal levels of crowding, however, benefits are quickly lost and various forms of pathology, occur. Growth may be stunted, life span shortened and various diseases or abnormal behaviors appear. Most studies on crowded animal populations in the last 20 years have concentrated primarily on the harmful effects of overcrowding. Such effects usually involve both physiologic and behavioral abnormalities.

The physiologic effect of overcrowding centre around endocrine imbalances of the adrenal pituitary system. These imbalances can lead to a great variety of disease condition including gastric ulcers, hypertension, nephrosclerosis, arteriosclerosis, and increased susceptibility of infectious diseases.

Prolonged exposure to crowding and its associated stresses have produced additions disease (hypoadrenalism or adrenal exhaustion) myasthenia (muscular weakness), and a variety of serious metabolic disorders leading to shock, coma, and even death. It should be pointed out that these stress effects are quite variable, however, and do not consistently reflected the same patterns in different individuals and various social groups. Behavioral pathologies also appear in experimental animals as a result of crowding and these are sometimes manifested more quickly than are the physiologic changes.

Overcrowded animals often show increased aggression and violence, abnormal sexual behavior, disruption of normal nest building and maintenance, breakup of stable social groups, disappearance of alteration of normal social roles, parental desertion of young and frequently cannibalism of young. The leading question, of course, is whether or not similar processes occur in man as a result of overcrowding. There are so many variables in man which affect health and behavior, that this becomes a very difficult question. There is no doubt that some striking correlations of disease and behavior are associated with urban concentrations.

Many human problems seem to be so accentuated in crowded populations that they virtually constitute an "inner city syndrome" disease, and emphysema; maternal problems such as high prenatal loss, parturitional difficulties and high infant mortlity; behaviour problems such as alcoholism, drug abuse, mental illness, and criminal assault; and social problems such as high divorce rates and social instability.

More recently it has been found that high prevalence of mental illness in the lower socioeconomic classes and dense urban ghettos of several other cities. The three volume report on Manhattan entitled, "Mental Health in Metropolis," also found unusually high prevalence of mental illness in crowded inner city inhabitants.

The problem of urban man seem so strikingly similar to those effects produced in overcrowded animal populations, it is tempting to conclude that similar processes are occurring. But, as stated previously, many factors are interacting with crowding in urban manpower, malnutrition, lack of educational and recreational opportunities, various environmental hazards, and unstable social patterns—that it is virtually impossible to separate cause and effect relationships. There are enough solid exceptions to the principle of "high density produces high pathology" that one must be cautious in carrying this dogma too far.

For example, the young nation of Singapore is obviously crowded—over 2,000,000 people on an island of 224 square miles, making it one of the most densely populated nations of the world—yet it enjoys remarkable good health. The death rate is one of the lowest in the world (5.5 per 1000) and the prenatal mortality is lower than that of Britain and the United States.

Major factors in this good record many be its youthful population (40 per cent of the Singapore population is under 15 years of age), and the well planned interspersion of parks and greenbelts with residential and commercial areas. One does not feel a "crowded" in generalizing as in Hong Kong, Tokyo, New York and Calcutta. Such variable demonstrate the difficulty of generalizing broadly about crowding or population density as a pathologic influence. Thus, the precise role of crowding in the "inner city syndrome" must remain a cloudily issue until more and better research studies are available. An encouraging development of recent years is the awareness on the part of some architects, city planner anthropologist and others of the relationship between our environment and mental health.

In Architectural Environment and our Mental health, Moiler explores the impact of physical space and structure on the physiological well-being of man. He noted that more buildings will be built in the next ten years than have accumulated since the beginning of civilization." And he wisely emphasized that, "We should recognize that if we continue to permit the structuring of a physical environment which is essentially hostile, to which man must somehow try to adopt, the cost to individuals and to society will be far too grat—in terms of mental illness, delinquency, poor motivation and fulfillment of our capabilities for creative work and community usefulness."

Anthropologists have also been interested in the spatial behaviour of man, and the book by Professor E.T. Hall entitled "The Hiddeen Dimension," was the first one to highlight some of the behavior patterns of man relating to social spacing. Hall showed the existence of "personal space" individuals and pointed out cultural differences in the extent and importance of personal space. A book edited by Esser reviewed the interest of psychiatrists, sociologists, biologists and psychologists, in the relationship of ecology and environment of human behavior. It emphasized how little we know about ourselves in these vital matters. Although most of the data on human spatial behavior are preliminary, it is a hopeful sign that several professional and scientific disciplines are now giving attention to this topic.

5 Consequences of Crowding

Crowding deals with a subjective psychological state when there are too many people in too little a space. The feeling of crowding is negative and has elements of stress. Typically one or many of the following characteristics are associated with crowding. Feeling of discomfort, loss of control over one's environment, lack of privacy, lack of initiative in soical interaction, fatigue, boredom etc. Crowding therefore, has perceptual affective and behavioural consequences. The outcome of such a phychological state is use of strategies that can reduce the feelings of crowding.

Crowding has two aspects to it.

1. *Social Density*: This refers to the number of people in a given space.
2. *Spatial Density:* This refers to the amount of space available for a given number of people.

Some recent studies of animal populations in which overcrowding is related to various behavioural pathologies are most suggestive of the stress produced by population density. Calhoun was one of the first to note that disturbances in the territorial and ecological distribution of animals may result in behaviour pathology and physiological malfunctioning. In mice, overcrowding in laboratory cages leads to abnormal sexual behaviour, decreased reproductive and nursing capacities, aborted pregnancies, deficient maternal care for the young and disrupted nest building. Calhoun reared a colony of Norway rats in four different pens, the rats tended to live together in one very-crowded pen.

Calhoun coined the phrase "pathological togetherness" to describe this behaviour which lowered the rats' fertility and shortened their lives. Other biologists have observed that mice crowded together in cages display a marked enlargement of the adrenals, a sign of stress; and there have been reports of a high incidence of cannibalism and male impotence in crowded mice colonies, adjustive mechanisms which function to keep down population numbers when they rise above a critical point.

It is not possible to dismiss these pathological reactions as unusual phenomena associated with artificial environmental conditions set up in a laboratory, for such behaviour occurs also in natural environments. Deevey has reported on the behaviour of lemmings, rat-like animals found in great aboudance in Norway.

Approximately every four years, when the population becomes excessive for the available territory, there is a mass migration in which huge numbers race blindly to the sea and drown. Such panic-like reactions to overcrowding in laboratory and natural environments are evidence of regulatory mechanisms that enable population control and, thereby, survival of the species. Whether extreme overcrowding in human populations produces analogous reactions of stress and social pathology is a moot-point.

Stimulus Overload Model

According to this model, high population density is a stressor as it stimulates the individual, excessively. Too many people in too little space not only crowd the environment about engage in more activities creating more noise, more pollution, more distraction etc. all of which overloads the system. This results in the inability to handle all the incoming information. A consequence of this condition is an aversive feeling. To reduce the overload of stimulation caused by crowding, individuals engage in behaviors that reduce the unavoidable interference from others.

At the physical level, the overload leads to fatigue and this is reduced behaviourably by withdrawing. At the cognitive level, since many cognitive stimuli, have to be handled, it results in confusion. Some investigators also regard excessively stimulating environments as causing stress by reducing the attention span, resulting in impaired cognitive functioning. Less friendly and less helpful behaviors seen among city dwellers is an example of such strategy used for coping with stimulus overload. In homes there is greater dissatisfaction among members and desire to spend time outside is more.

Behavioural Constraints Model

High density conditions create stress by imposing restrictions on behavioural freedom. This limitation can produce a psychological reaction whereby the individual is motivated to remove the restriction by either withdrawing or by compensating for the density by engaging in behaviours that help gain control over the environment. For example in a crowded home or hostel people delineate area or seats for themselves (my chair, my corner, my window etc.) as a means of gaining some control over one's crowded environment.

Ecological Model

High density conditions lead to disruptions in accurately understanding the functions of objects in terms of their affordance. All objects in the environment, convey affordances about their functions. Eg. A fruit says eat me; a ball says kick me; a baby says protect me etc. These affordances act as cues for adapting to the environment, efficiently. High density environments block these affordances resulting in the misreading of the cues regarding their functions.

For example, a family which should be perceived as supportive and emotionally positive, due to population density is seen more as a nuisance, rather than a source of joy or support. In such a perceived situation the absence of close-warm feelings among members and the desire to form smaller cliques etc. are easily understandable. Many extended families are now breaking up into separate entities or becoming fully nuclear one of the reasons for this being crowding.

The situation is further compounded by the scarcity of social roles and physical resources. For eg. The grand father plays the senior role member, even though the adult sons have taken over the functions of providers for the family. The thwarting of such roles and privileges can lead to a feeling of lack of control over one's environment. Behavioural consequences of this would be to opt for situations where control over ones environments possible, failing which, the stress is experienced without relief. This can give rise to negative feelings about the family.

Attribution Model

In this model individuals who experience lack of personal space due to crowding and therefore feel the negative attributes the stress to space encroachments. Eg. When too many guests arrive and space is short, there is sense of relief when the people leave. However, there are certain situations when high density is not very negatively perceived Eg. An enjoyable and voluntary party. So the critical element in response to limitations of space lies in attributing the negative feelings to crowding. If the attribution is absent; then the sense of crowding is not experienced, so negatively, even though the environment is a high density one.

Stress and Arousal Model

How do high density conditions produce stress? Any physical, social or emotional stimuli that are unsettling can be described as stressful. Stressful conditions are characterized by:-

a) Stimuli that are very intense.

b) Beyond the limits of one's control.

c) Threatening, due to one's inability to cope with it.

High density environments span across these three factors making a stressful situation.

Sources of stress have also been identified:.

1. *Physical*: Any intense stimulus arouses stressful psycho-physiological responses. Extreme noises, temperative pain etc. are conditions that cause stress.
2. *Psychological*: Deficiencies in personality make up stress the individuality impeding the progress towards the goal. Eg. Inadequate skills, attitudes, poor leadership qualities, having an avoidant personality etc.
3. *Social/Environmental*: Social factors like collective norms traditions etc. can stress the individual by acting as barriers to goal attainment. Environmental factors that stress the individual include, noise, pollution, poor services, over crowding etc. Stress involves s situation where the individual experiences threat to his important goals and also feels that he is unable to cope with the threat. There is also an impending sense of going to be overwhelmed.

Stresses can arise from major traumatic life events like death of a spouse, parent, personal injury, retirement, pregnancy etc., or from the frequent minor hassles of everyday living. Fortunately traumatic events though extremely stressful, are rare. The minor daily stresses make up for their low intensity by their nagging frequency.

Stress affects the physical and psychological health of the individual. It has implications for performance in tasks as well as on social behavior. The effects of stress on physical health has been well researched. Stress is thought to be silent killer. Medical reports suggest that stress affects health to the tune of 50-70%. Stress has been implicated in the occurrence of diseases like heart attacks, hypertension, hardening of the arteries, ulcers and even diabetes. This is because stress renders the immunological system inefficient. The more prolonged the stress is, the greater is the threat to the immune system of the body.

Studies done on individuals who are in mourning or about to take an important examination have been found to show disruptions in their immune system. However, what is encouraging is, that once these high stress levels are lessened, the restoration of the immune system, is rapid. Stress inevitably arouses the organism. This is a survival oriented reaction of the organism to meet the emergency situation induced by the stress. The body's resources are then mobilized to meet the threat. This is a demanding and exhausting process. Therefore, if stress is constant or/frequent it takes a toll on the physical health and psychological well being of the individual.

Crowding qualifies as a constant stressor. In order to escape the harmful effects, the individual engages in behaviours that seek to lessen the density and thereby the experience of crowding. This in turn would lead to a lowering of the stress. Typically this manifests in the form of less helpful and less friendly behaviours. This can also explain the urban/rural divide in terms of friendliness and feelings for others. The long term effects of crowding have shown higher arousal of the system. Arousal is a form of energization of the system. The greater the stress, the greater the arousal. However, adaptations to the stress also occur. But the adaptation generally involves raising the base line. This means that the threshold for stress tolerance becomes lower. Subsequently in real terms it means that events produce less stress are also, now capable of effecting the organism.

Since the effects of high population density are generally negative, it has consequences for social behaviour. In a study, people who were already in a crowded room were informed that more people are going to come in. The anticipation of more crowding led to feelings of discomfort. It showed up in reduced liking of even fellows waiting alongside. This is seen in a train compartment. Even as the train enters a station the possibility of more people entering produce stress. This leads to overall reduction in friendliness. Significant reduction in helpful behaviour like making room for luggage or staying clear of the passage etc. can be explained in terms of the effects of the crowding.

Individuals who are required to be in environments that are both confining and at the same time be constantly exposed to other individuals as in prisons and other settings like remand homes, free hostels etc. showed greater irritability and reduced satisfaction and harmony in social relationships. In the community settlement the "Kiffutz" in Israel, where children are raised in a community with multiple mothering lacked privacy. This absence of privacy and constant exposure to others resulted in feelings of tension. Anxiety and aggression were some of the other characteristics that were observed. Students living in three and four seater rooms expressed greater dissatisfaction with their roommates as compared to those living in two seater rooms. The latter also felt less control over their interactions.

Some studies have shown that the grades obtained by the students in the more crowded conditions were lower than those from the more comfortable settings. These differences can be attributed to social density. Behaviours that follow, typically involve attempts to reduce the stress and unpleasantness brought about by crowding. Withdrawing and/or less friendly/helpful orientations are the most common forms of response.

Table below shows the satisfactions derived from roommates in two conditions.

Satisfaction with Roommate under crowded and uncrowded conditions

	Uncrowded	Crowded
Satisfaction with roommate	4.9	3.7
Perceived cooperativeness of roommate	4.7	3.9

The lowered frequency of interaction in high density places can be seen as the effects of less attraction for neighbours. This can then be attributed to crowding whereby lessening of contacts is a way of ensuring reducing the unpleasantness of crowding. Although it appears that high density leads to lower attraction, there is evidence that under conditions of high spatial density, this response is more characteristic of males than of females. In an experiment male and female subjects participated in same-sex groups of six.

Consistent findings on a variety of measures indicated that while males responded more negatively to group members in high than in low spatial density conditions, females, liked group members more under high density conditions. They state that while both males and females experience arousal from high density, social norms permit females to share their distress with others in their group, which leads to liking and cohesion. The same norms prohibit males from sharing stress, which causes a more negative response. In a follow-up experimental, 1976, it was found that when females were not permitted to interact with each other, their positive reactions to high density were limited.

Prosocial Behaviour. If high density leads to lower attraction and to withdrawal responses, how might it affect helping? For example, suppose you lost a contact lens or something else of value. Where would you be most confident of finding someone who would help you look for it; in a high-rise or a low-rise building? In a large city or in a small town? Interestingly, most research on how density affects prosocial behavior has been done in field setfings like these. In fact, one set of studies has compared helping in building characterized by high and low density, and a second has compared prosocial acts in large cities and small towns. In studies that explored helping as a function of building density, it was found consistently that greater density leads to less helping.

Studies compared prosocial acts in high, medium, and low density dormitories. Envelopes, which were stamped and addressed, were dropped in the dormitories, and helpfulness was measured by the number that were picked up and placed in the mail. The results showed that 58 percent were mailed in the high density condition, 79 per cent in the medium density condition, and 88 per cent in the low density condition. In an interesting corroborative study, the effect of social density on compliance with a prosocial request (printed on signs) for cafeteria users to return their trays to designated areas. It was found that fewer users complied during high density periods.

The second line of research, which compared urban and rural helping, provides more equivocal evidence concerning a density-prosocial behaviour relationship. In one group of studies it was found that city dwellers were less helpful and less informative to strangers than rural dwellers. Urbanities were less likely to admit a stranger who needed to use a telephone, less likely to assist a caller who had mistakenly called a wrong number, less likely to correct over payments in making change, and less likely to report theft than rural dwellers. A number of interesting field studies have shown that resident of low-rise buildings are more willing to help others than residents of high-rise buildings.

Withdrawal. Withdrawal may be associated with high levels of social contact. Studies have found that withdrawal may function as an anticipatory response to high density, as a means of coping with ongoing high density, and as an after effect. For example, the mere expectation of high social density elicits withdrawal responses, including lower levels of eye contact, head movements away from others and maintenance of greater interpersonal distances. Withdrawal also occurs during ongoing high density interactions: Subjects are more willing to discuss intimate topics under low density conditions, and both children and psychiatric patients interact less frequently as rỏom density increase. Finally, it appears that withdrawal can constitute an after effect of exposure to density. Males were less likely to volunteer for another experimental session after high social density and groups of males preferred larger personal space and recalled fewer names after exposure to high density.

Aggression: If density can make people less helpful or want to withdraw; then it is possible that density can also lead to hurting behaviours. Many studies have also established greater aggression and hostility from high population areas. Apart from crowding the high density areas like the slums, require that people have to compete for scarce resources also. For eg. space, water etc. Hostilities can break out frequently in these conditions because they have conditions which are highly tension producing. Competition is a major possible source of interpersonal conflict. This conflict can easily become inter-group conflict under conditions that threaten limited vital resources. These would be heightened by feelings of insecurity due to dissatisfaction of needs, resulting from population pressures.

Mean Scores on the Feeling of Crowding and Social Behaviour for Outside-Inside Density Groups

Groups		*Mean*	*df*	*t*
Outside	High density	61.00	98	49
	Low density	60.00		
Inside	High density	89.00	48	7.49
	Low density	46.70		

One study is tallisled that moderately high density conditions led to increased aggression in males, while very low and very high density conditions led to decreased aggressiveness. In another study, aggression was found to increase in boys under conditions of high density. These findings can be explained in terms of competition for limited resources which males, more aggressive as a group.

Sex differences have been observed with respect to aggression. Same sex groups of eight members were placed in a room that was either large or small. Males rated themselves as more aggressive in the small rooms as compared to females. Studies of boys and girls in the class room as well as in anticipation of more cordoned, produced more aggression among males. Spatial density, rather than social density seems to have more aggressive outcomes from males. Inside and outside conditions of population density and its effects on social behaviour were studied. It was found that density is a necessary but not a sufficient condition for producing feelings of crowding.

Social Behaviours and the Feelings of Crowding Experienced by the Residents of High and Low Inside Density Conditions

Behaviour/Feeling	*Density*	*Mean*	*t(48)*
Crowding	High	3.55	3.33
	Low	1.82	
Privacy	High	3.07	
	Low	1.38	3.07
Physical condition	High	3.22	
	Low	1.44	4.45
Interaction	High	3.24	
	Low	1.38	3.20
Neighbours	High	2.65	
	Low	1.69	3.92
Insecurity	High	2.67	
	Low	1.24	6.5
Illness	High	3.44	
	Low	1.71	3.39
Withdrawal	High	3.88	
	Low	2.46	2.43

But high density within the house had significant effects on the feeling of crowding. The following talk gives a comparison of the behaviour feelings of individual living in high low density conditions. Further, crowding affected health, feelings of security, privacy and generally produced the desire to leave. Negative moods and dislike for affiliation were also related to crowding inside the house. The feelings of crowding in high and low socio-economic studies groups have been compared, and are given below:

The Feeling of Crowding in High and Low Economic Status Groups

Dimensions of the Feelings of Crowding	*Economic Status*	*mean*	*t (48)*
Comfortable-uncomfortable	High	1.97	
	Low	2.34	2.64
Good-bad	High	1.98	
	Low	2.48	3.57
High Congestion-low congestion	High	2.02	
	Low	2.14	0.32
Happy-sad	High	2.08	
	Low	2.48	2.85
Pleasant-unpleasant	High	1.86	
	Low	2.40	4.00
Positive-negative	High	1.86	
	Low	2.36	3.57

The following table shows the feeling of crowding under high and low density conditions in terms several dimensions.

The feeling of Crowding in High and Low Density Groups

Dimensions of the Feeling of Crowding	*Density*	*Mean*	*t (48)*
Comfortable-uncomfortable	High	2.32	
	Low	2.00	2.28*
Good-Bad	High	2.40	
	Low	2.04	2.57*
High congestion-low congestion	High	2.00	
	Low	2.14	0.81
Happy-sad	High	2.38	
	Low	2.18	1.42
Pleasant-unpleasant	High	2.28	
	Low	1.98	2.14*
Positive-negative	High	2.25	
	Low	2.10	1.07

The Jain findings show that high density in itself does not lead to low degree of attraction towards neighbours; but high density conditions can aggravate aversive feelings if space and resources are extremely limited and have to be shared. This finding has implications for architectural and colony features that have to be kept in mind while designing high density housing localities.

In the Indian societal conditions where emotional bonds between relatives, neighbours and community have been traditionally emphasised; the western models which predict low inter personal attraction in high density areas, cannot be fully observed. In fact it is, the high socio-economic class who are also highly competitive and show less interpersonal attraction for neighbours and others, as compared to the lower socio economic group.

The typical helping out relatives, neighbours etc. in times of funerals, weddings etc. is more characteristics of the highly populated, lower economic group. The results thus show that economic factors interact with high density and societal values to moderate the effects of high density living conditions on social behaviours.

Effects of High Density on Task Performance

One of the most critical questions that can be asked about high density is whether it affects task performance. The answer has important implications for the design of all types of living and working spaces (e.g., schools, factories). Most early studies used simple tasks and were consistent in finding no decrements under high social to spatial density. As seem in the table below both high social and high spatial density led to decrements in complex maze task performance, but these decrements were more pronounced under conditions of high social density. In a field setting, also decrements in complex task performance over time in overcrowded dormitory rooms (three persons in a room built for two) were observed.

Errors in Maze Performance as a Function of Spatial Density and social Density

Low spatial density	34.20
High spatial density	37.44
Low social density	32.13
High social density	39.50

In another study poorer complex task performance under crowded conditions was noticed, but there was no impairment in simple task performance. Even anticipation of crowding led to poorer complex task performance. The differential effects of high density conditions on single and complex tasks can be explained using the concept of arousal. Arousal when moderate has the properties of being a motivator. But when the organism is over aroused as in the case of overcrowding, the anxiety is experienced.

Anxiety is detrimental to learning & performance. Complex tasks which need greater attention & problem solving capabilities is more adversely affected than simple tasks. So high density is more adverse for complex task performance than simple tasks. When high density conditions prevail & the expectations of task performance are also low; it yields significantly poorer functioning.

In another study interaction among members was found to be an important factor that produces lowering of tasks performance in a high density condition. A final and very important question is whether high density can cause aftereffects, as well as immediate effects, on performance. As you will recall, noise is a stressor that has been linked to consequences occurring after exposure, and some studies suggest that crowding has lingering effects as well. For example, it was found that subjects exposed to high density later showed less persistence at working on unsolvable puzzles than subjects exposed to low density.

In an attempt to determine whether perceived control would lessen affereffects, some subjects had the option of leaving a crowded room to complete the study in a spacious setting (i.e., perceived control). Although no one took advantage of the option, this group showed fewer aftereffects than the group that wasn't offered an opportunity to leave. The similarity of these findings to those with noise effects further suggests that noise and high density may affect people in similar ways.

Performance suffering under high density, can also be explained as stress induced. Stress has been found to systematically lower performance primarily because it distracts the person from concentrating on the task on hand. Lack of concentration in turn is because of the unpleasantness of the experience. This shift in focus from task to the negative emotions reduces efficiency. Complex tasks stand to be more readily disrupted by stress as these require greater concentration and therefore suffer larger deficits. Decision making under conditions of stress have been described as resulting in bad choices. The strategies utilised in problem-solving under stress conditions have been found to be poor.

Many premature decisions are also associated with choices made under stress. Scanning of the possible alternatives for making decisions has been found to be disorganized under conditions of stress. A tendency to close the issue on hand as quickly as possible is linked to the presence of stress. Here the individual is motivated to overcome the aversiveness of the stress and therefore quickly engages in actions to reduce it. Premature closure without considering all aspects of a given situation results, owing to the stress effects. Crowding not only induces stress and thereby lowers performance but can also be implicated in poorer decisions and premature closures. Crowding also introduces the stimulus overload condition. To reduce the overload, individuals engage in hasty and ill-conceived decisions. Moderate levels of stress which have motivational properties of energizing the organism, has significant positive features, in terms of performance.

Medium amounts of stress are considered as arousing the organism to better performance through increased mobilization of energy and better attentional focus. However, when they cross the optimal conditions, then the effects are negative and adversely affect performance. Crowding, due to its excessive stimulating features, qualifies more as a stressor and thereby has deleterious consequences on performance. The only kind of skills that are not very adversely affected are the well established ones like driving, swimming, playing an instrument or speaking a language. Crowding as a stressor, has significant effects on the health of the individual. This in turn has negative consequences for learning & performance.

Since the effects of crowding on health is also infavourable, performance can also suffer on account of reduced health in an indirect manner. The effect of stress on learning is even more disruptive. Learning involves the acquisition of a new set of responses. This implies that the correct response exists at the 0% level of probability of occuring. This low probability response is at the bottom of the hierarchy. Learning means pushing this 0% probability response to one of 100% probability. Such an endeavour requires high degrees of attention and cognitive imputs. Stress induced by crowding, arouses the organism and such an arousal state is energizing.

When learning is engaged in this energized condition, the first natural response to occur is a previously learnt 100% probability one. The new response (to be acquired) has to now compete with the well established earlier response. By the very nature of the process, the newer response has a disadvantage and would therefore suffer, seriously. Research evidence shows that even the mere presence of others is deleterious to learning.

Crowding (which is the presence of many people) would lower learning not only because of the presence of others, but more so because of too many people. Thus, stress induced by crowding would hamper learning in an even more pronounced manner. The new responses to be learnt would in the competition with the older established ones, get depressed easily. So greater effort and attention would be required to offset this condition. Education imparted in high density classrooms is found to be ineffective. Further, children who study in classes with large numbers, are also more likely to come from homes that are crowded, than otherwise. This would suggest that both the conditions are unfavourable for promoting learning or that extra efforts would have to be made for reasonable levels of learning to occur.

In such strained conditions, students would also not readily have role models of sustained learners. To compound it, other stressors would be in operation in such disadvantaged environments. All these together pose a set of factors that are not conducive to serious learning activities. This could well be the explanation for the absence of children from the slum sections benefitting from long-term education.

Crowding inevitably leads to greater competition for both space as well as resources. If competition is continuous, it can produce increased tolerance for competition according to Uday Jain, who has done pioneering work on crowding in India. Scarcity of resources typically manifests in struggle for power with a view to cornering the scant resources. In India, the immense desire for power is not so much for power itself but more because power symbolises access to resources. Obtaining railway tickets, getting admissions in educational institutions, jobs, telephone connections, licences for business and industry, etc., are all related to power as a means of need-gratification.

In Jain's study it was clearly demonstrated that high density situations lead to greater competitiveness as a way to obtaining the limited available resources. Urban people from high density areas showed more competition tolerance than those from the low density areas. Similarly, the high density urban individuals competed more than the high density rural individuals. High and low density areas in the rural setting also yielded similar results. In this situation of competing for scarce resources, people hardly find time to think or be mindful of others' needs. This can account for the apparent indifference of the present day city dweller.

Given that one has limited time/energy, etc. for obtaining some benefits for oneself, expecting one to be very caring and considerate to others seems to be a very unreasonable one. The blanket statement made about the younger generation perhaps needs to be reinterpreted using the concepts of high levels of competition generated by high levels of population. It is a clear that the earlier generation competed less for the resources as there was a larger amount that could be shared. If competition becomes fierce as the population increases, it is possibly that the individual's interpersonal relationships would be markedly affected. The biggest causality of such a condition would be consideration for others.

Studies show that there are urban/rural differences in the extent to which they tolerate competition. Greater tolerance for competition has also been associated with the need patterns of the high and the low competitive individuals. Low competitiveness is associated with helpful and considerate behaviours while high competitiveness is related to order, neatness, organization, etc.

The latters needs are likely to be more readily disrupted in a crowded environment. Hence the reduced consideration and helping behavior found among the high competitive individuals can be easily attributed to their negative perceptions of others who are viewed as rivals. This is due to the fact that too many persons would not only increase the competition but also intrude into their neat and orderly styles of functioning. Such a situation would naturally yield less consideration and helpful behaviour. High density dwellers have to encounter vast numbers in their daily activities. They try to conserve their psychic energy by keeping interaction to the barest possible number of people. In such a context, friendly, helpful and considerate social interactions are highly unlikely. This is more clear among the more educated who are more competitive and thus uninterested in nurturing supportive behaviours with others with any degree of spontaneity.

So, the explorations of several density-behaviour relationships shows that high density has various kinds of negative effects. High density is like any other potential stressor and it has

(1) Immediate effects, such as physiological arousal and negative states.

(2) Coping responses like withdrawal, aggression etc.

(3) After-effects and cumulative effects like health and performance.

(4) Motivational consequences (increased tolerance for competition)

Summary of Theoretical Perspectives on Crowding

Conceptual Approach	Critical Causes(s) Crowding	of Primary Coping Mechanisms
Social overload	Excessive social contact too much social stimulation	Eacape stimulation; prioritize input and disregard low priorities withdrawal
Behaviour constraint	Reduced behavioural freedom	Aggressive behaviour leave situtiom coordinate actions with other
Unwanted interaction	Excessive unregulable or unwanted contact with others	Witdrawal; organization of small primary groups
Interference	Disruption or blocking of goal-directed behaviour	Create structure aggression; escape
Ecological	Scarcity of resources of outsiders	Defense of group boundaries; exclusion

All these models are overlapping and are supplementary in terms of explaining the stress experienced crowding and the behavioural consequences that follow, being stress-alleviating in nature. Recent conceptualisations on the effects of high density revolve around two concepts control and privacy. Perceived control is regarded as a potent mediator of stress. When we believe that we can control a stressor or other aspects of a situation, the aversiveness of stress appears to be reduced. It follows that even if no other problems are apparent, losing or not having control can also be stressful. Several researchers have propsed that high density can cause a loss of control (or prevent some one from ever having control) and that this loss of control is the primary mechanism by which density causes stress. Can we really explain the negative effects of high density as a loss of control? The answer is "Yes," even though at times the explanation seems a little strained. Some explanations fit well the notion of unwanted interaction is a control-based perspective. Negative effects are the result of contact that is too frequent which makes control over when, where, and with whom people interact difficult to manintain. As a result, interactions become unpredictable and frequently unwanted.

The behaviour constraint notion is also control-based, decribing high density as eliminating behavioural options and reducing freedom to behave as one might like. For example, having inadequate space can constrain our behaviour by making it impossible to control the nature of interaction with others. We can generally control the degree of intimacy in one-on-one interactions by adjusting the distance we stand from people but in a very crowded room we may find ourselves with no choice-we must stand close to people whether we know them well or not. In this manner, we can lose control over intimacy regulation.

Interference can also be viewed as a threat to control as our attempts to achieve one or another goal are repeatedly blocked or disrupted. And resource problems the focus of ecological models, can limit our choices and restrict our ability to exercise control. To some extent, all of these consequences of high density that have been related to negative effects do cause a reduction in control. But, is there any direct research evidence for the relation between density and control?

Research examining the links between high density and loss of control has taken two different tracks. The first has been to manipulate personal control (provide some subjects and not others with perceived cantrol) and see if it has any effect on experience in high density settings.

Example subjects response to riding in an elevator crowded with confederate of the experimenter. They observed a tendency for people to gravitate toward the elevator floor selection panel (a control panel, if you will), so they manipulated whether subjects were able to stand near it (high control condition) or not (low control condition).

The results indicated that subjects standing near the panel felt better than those in a defferent corner. They also thought that the elevator was larger than subjects not near the panel. A some what different approach found that noise was associated with negative after effects which was reduced when subjects had some degree of control over the noise.

A similar study with density instead of noise was carried out. Once again, negative after effects were associated with stress when control was not available. Subjects who had more perceived control, on the other hand, did not exhibit negative after effects following exposure to high density.

In a high density condition in the laboratory, subjects had varying degrees of control over the setting. Those with control felt better than those without. These studies demonstrate that control can reduce the effects of high density relative to no-control, high density experiences. They do not, however, demonstrate that high density itself has consequences similar to those associated with loss or lack of control.

However, studies conducted in field settings found that chronically high residential density was associated with helplessness-like behaviour. Learned helplessness is a syndrome in which people who are exposed to uncontrollable setting learn that they cannot control the setting, and hence stop trying to do so. This manifests itself in reduced motivtion and congnitive activity. Subjects, who were children and adolescents showed symptoms of helplessness that were associated with density in their homes. Symptoms of helplessness were also found among people exposed to high density in their residential environment over prolonged periods. More recently this has been directly linked to perceptions of control in residential settings. As people relinquished their beliefs that they could control these settings, their behaviour became increasingly like that asssociated with helplessness. All of these studies taken together suggest that control is involved in negative high density effects. When control is available, high density has less impact on pepole than when it is not available. Further chronic exposure to high density appears to be asociated with learned helplessness.

Imagine yourself growing up in a small apartment with five other people who are continuously interacting with each other and with you. With so many people in so little space, you may grow up to feel the world is a complex place in which you have little power to influence events (What consequences does this have? When we come to believe we cannot control our outcomes by responding appropriately (as may result from living in high density conditions), we no longer perform effectively in a number of situations. This syndrome is called "learned helplessness." Two interesting and provocative studies demonstrated the relationship between residential density and susceptibility to "learned helplessness".

In one experiment it was found that children who lived in high density conditions were less likely than those in low density to try to control the administration of rewards they were to receive. In a second experiment subjects from both groups to an initial frustration task on which responses and outcomes were noncontingent children from high density homes did significantly worse on a subsequent task for which outcomes were contingent. Thus it appears that home density is an important determinant of both the use of control and performance after frustrating noncontingent reward situations. Besides the home environment what other settings may be source of 'learned helplessness' training? One important culprit may be the schools.

As classroom size increases, learned helplessness training begins to occur. They argue that larger classroom sizes lead to lower student expectations for control of reinforcement, because teacher feedback concerning student work becomes less descriminative. For example, as class size goes up, individualized student-teacher interactions decrease, and generalized (rather than individualized) praise and criticism increase. Clearly such conditions could lead to a state of learned helplessness with its negative consequences for performance. This could have profound consequences for learning. An alternative to the density-intensity model has been proposed by Freedman. He agrues that density only intensifies reactions that would occur in any case in a particular situation.

High density according to him, heightens the importance of other people and magnifies our reactions to them. Thus high density would intensify the aversiveness of an unpleasant situation and intensify the pleasantness of positive situations. From this point of view, a number of factors can lead to a negative reaction in a high density situation. One of the benefits of such a model which utilises the notion of intensfication; is the link between crowding and contagion. Contagion is the rapid spread of behaviours or emotions from one person to the group. Eg. People coming back after viewing a game. The negative effect, (if the home team loses) would be quickly passed on among all those returning.

This model provides a frame work that specifies three factors as important in determining the experience of crowding. They are:

(1) Individual difference

(2) Situational conditions

(3) Social Conditions

Identifying individual difference variables can help determine who among the selected persons will be more sensitive to the constraint of space. Studies have already established gender differences in the aversire feelings, due to crowding.Some persons prefer to sit for away from others while others prefer to sit closer. This can also affect the degree to which crowding is experienced.

Personality characteristics also moderate our reactions to high density. It has been found that internals (who feel they control their fate) display a higher threshold of crowding than externals (who feel events are controlled by outside forces). The primary problems encountered in high density settings may also affect the ways in which individual differences are important. People who screen themselves from interaction and organize their surroundings were better able to cope with social density than were people who did not screen themselves. One would expect that this variable would be less important when spatial problems were paramount.

A final individual difference variable that has been linked to reactions to high density is one's adaptation level (from past experience under high density conditions). A number of investigators have hypothesized that people with a history of high density living are less likely to experience crowding in a novel situation than are people with a history of isolation. For example it has been found that the Japanese, residents of HongKong and the Logoli (all of whom live under extremely high density) have developed social mechanisms that may be viewed as adaptive for high density living.

In addition to being moderated by individual difference variables, our reactions to high density are also affected by *situational conditions*. An important situational condition is the degree of control we have. Complementing those discussed earlier, several additional studies have found that allowing people increased control over a situation, leads to less perceived crowding. The applied potential for introducing control in high density conditions, is phenomenal. For e.g. Individuals living in high denity situations, can be given training for enhancing control. ie. how to share space and ensure privacy. This is bound to help alleviate crowding and it negative effects.

Other situational factors as whether we are in a primary environment (like a home) or a secondary environment (like a restaurant), and the extent to which other stressors, such as noise etc. are involved, also affect crowding. When we are in a primary environment we may be more likely to experience crowding. Finally, in terms of activities there is suggestive evidence that we are more likely to experience crowding when engaged in work, rather than in recreation. Life styles that increase social contacts like picnics, games, travel etc are more likely to generate feelings of crowding in dense environments than life styles that decrease the likelihood of social contacts. In addition to individual differences and situational conitions, *social conditions* can be manipiulated to affect whether or not we are crowded. These are variables that make up the social "climate" of a high density situation such as the degree of friendship and the level of social interference and interaction.

For example our relationship with the people were with, may determine how crowded we feel: Less crowding is experienced with liked rather than disliked others and with acquaintances than with strangers. To the extent that we experience social interference (interruptions) by others or excessive proximity or intimacy (too direct eye contact or body orientation) crowding is more likely. Finally crowding is more often experienced in unstructured task situations than in structured ones. High density within primary groups (e.g. families) has repeatedly shown fewer efects than it has in other groups—when density within apartment units in examined, for example it appears to be negligible as a factor associated with illness or behavioral difficulties.

Degree of acquaintance and relative position of an individual in a group's dominance hierarchy also affects crowding. The presence of friends or the possession of high status tends to blunt the aversiveness of large numbers of people or cramped spaces. Since some of the conditions and factors that mediate the experience of crowding have been delineated; it is worth considering how one can modify existing environments or plan new ones to mitigate the effects of crowding.

To begin with, greater ceiling height is asssociated with a feeling of less crowding. Rectangular rooms seems to elicit less crowding than square rooms, of the same area. Rooms that have visual escapes such as doors or windows are rated less crowded than rooms that do not have such escapes. The adding of flexible partitions in rooms, lessens feelings of crowding. Increasing the brightness of rooms by providing suitable wall colors leads to less perceived crowding.

The presence of visual distractors e.g. pictures, some movement etc. also helps reduce crowding. Finally, studies that have evaluated the aversive effects of Social *vs* Spatial density have established that social density is more negative than spatial density. Architects and designers must address themselves to the fact that too many people produces greater sense of crowding than too little space.

Cities are the hotled of crimes of different kinds owing to a peculiar mix of people, social, physical and housing factors. Large scale rural to urban migration; the concentration of the underprivileged in core unban areas, crowding of population into higher and higher densities all significantly affect life styles environmental conditions and crime rates, in the large metropolitan areas. It is now becoming increasingly clear that the physical and architectural designs of residential environments can elicit attitudes and behaviour from residents which contribute in a major way towards insuring their security Residential environments with certain physical features have crime-inhibiting qualities.

As a starter, 'semi-private' spaces should be 'designed in' to facilitate social interaction between residents and thus encourage the development of social order. Such areas should exist both inside and outside the building and should be planned so as to make pleasant social interactions inevitable. This is specially true of many South Asian countries and particularly India, where dependence and interdependence rather than independence is a valued social goal in interpersonal relationships. Many homes have extended families, which involve the living together of many individuals of different generations. So our housing colonies should be totally designed to suit our socio-cultural environmental therefore needs. So houses that alienate cannot be appropriate for India. Yet, colonies that grow up where spaces are 'nobodys' can become dumping grounds easy for encroachments or neglect and subtly support crime and randalism. In order to take care of this situation the design of a 'defensible space' has been proposed. This notion implies that building should be so designed that all the space appears to 'belong' to some individual or group. Such areas can be easily depended and therefore promote surveillance and thus reduce unauthorised activities and crime.

In practice this means that areas around buildings e.g. service lanes, parks playgrounds etc. must belong to the building. Then these areas would be perceived by both the residents of the building and outsiders as extensions of the building. The otherside of this issue is that public areas such as hallways, corridors, stairways etc. should be eliminated as far as possible. This can be done through horizontal rather than vertical constructions. Where this is not possible then open 'air spaces' like courtyards combined porches or hallways that are accessible to only a small group of residents should be constructed. These designs would facilitate territorial functioning. If corridors have to be provided; they can be made to run along the window side. This is easy to be monitored.Designs should also help police surveillance. Often buildings are not open to police viewing or even reaching at times of need.

6 ECOLOGY AND MAN

The most difficult and essential problems for man's future lie in broad ecologic and social issues. No two questions are more fundamental or more troublesome to the future of man than the ecologists' inquiry. The carrying capacity, or life support capability, of the earth for man may be dependent upon a great variety of factors. We can readily identify oxygen water, food space and shelter as vital requirements for human life. There might be many others as well such as mineral resources, energy availability, or the concentration of toxic metabolites, which could reach states above human tolerance levels. Most of these however, would operate through the media of air, water or food.

The atmosphere of the earth consists of about 78 percent nitrogen, 21 percent oxygen, 0.03 percent carbon dioxide, and the remainder, other gases. This composition is remarkably stable despite the fact that oxygen is a highly reactive element which readily combines with a great many other elements in chemical processes of oxidation. The ecologist Lamont C Cole has pointed out that oxygen has remained stable in the earth's atmosphere because green plants recycle it in molecular form through photosynthetic process. Approximately 70 percent of the free oxygen produced each year is through to come from planktonic plants in the ocean, and the remaining 30 percent from terrestrial vegetation of forests and grasslands. Cole emphasized that if we should seriously pollute the sea, as is already happening in coastal areas, and if we should excessively denude the forests and grasslands of the earth, we shall be in danger of upsetting the oxygen balance in the atmosphere.

In other words, we must look upon the oceans, forests and grasslands, as the world's oxygen tank, and we must insure their continued healthy ecologic function. This point of view is contested by W.S.Broecker, a geochemist at Columbia University Dr. Broecker points out that each square meter of the earth's surface is convered by 60,000 moles of oxygen gas.

Plants in the ocean and land produce only about 8 moles per square meter per year, an this is almost entirely consumed by their own respiration and that of animals and bacteria. Thus other organisms use all but 0.0004 percent of the oxygen produced by plants. Broecker maintains, therefore, that man does not depend upon the oxygen generated by plants, and that there is no current danger at all to his oxygen supply. "In conclusion it can be stated with some confidence that the molecular oxygen supply in the atmosphere and the broad expanse of open ocean are not threatened by man's activities in the foreseeable future.

Molecular oxygen is one resource that is virtually unlimited". Broecker feels that man's existence will be threatened by other types of environmental pollution long before his oxygen supply becomes endangered, and that we need not worry about total oxygen supply. This is typical of some of the different opinions which are unfolding between physical and biological scientists. Such differences often indicate the need for more research and better knowledge before we can fully accept one position on another. Another atmospheric problem of long-term concern is the increase in carbon dioxide in the atmosphere. Between 1860 and 1960 the combustion of fuel increased the CO_2 content of the air by 14 percent of its formerly stable level, Carbon dioxide tends to reflect infrared rays from the earth, and this provides a "greenhouse effect" for the whole earth; that is, it accumulates heat. This will gradually increase the temperature of the earth—an effect which may have long-term implications in climate modification.

A rise of just 2 or 3 degrees in annual mean temperature would have a major melting effect on glaciers and ice caps, and this could raise the level of the seas several feet. Such a rise in the seal level would inundate many of the world's major cities. Some experts predict it may require 4,000 years, and others feel it will not happen at all. The latter believe that any occurrence of the greenhouse effect due to increasing CO_2 in the atmosphere will be offset by increasing particulate pollution in the air and a cooling of the earth's surface.

Once gain, we see fundamental controversies on major issues-largely because we do not really have the complete answer. The atmospheric problem may also become acute in terms of various toxic elements released into surface, air and water: carbon monoxide, sulfur dioxide, lead from automobile exhaust, hydrocarbons from industrial operations and numerous other products of modern technological societies. For example, 500 million pounds of lead are discharged into the atmosphere of the United States each year from automobile exhausts alone, and significant amounts of this become incorporated into living organisms.

A recent study in Sweden showed concentrations of lead in mosses to be as high as 300 to 500 ppm along roadsides. Some roadside plants in the United States may have lead levels up to 3,000 ppm,. This lead enters the plants via both air and water and can then enter man and animals. A hundred years ago, before the automobile, concentrations in roadside plants were less than 20 ppm. Lead levels over 0.5 ppm have detectable physiological effects in man and animals, and 1.0 ppm levels are often toxic.

There seems to be little or no immediate threat that we will run out of air for breathing, but several warnings have appeared regarding the quality of our air. We apparently have an adequate supply of total oxygen for many centuries to come though even this is a debatable point. In some localities the situation may be critical at certain times. If stationary inversions of air trap the polluted atmosphere over some of our largest cities such as Tokyo, London, New York or Los Angels, toxic air pollutants increase to dangerous levels.

In August of 1969, a massive still-air inversion occurred in eastern United States, causing serious air pollution in urban areas from New Orleans and St. Louis, north to Chicago and Milwaukee, and east to Washington and Baltimore. At no time was there an absolute shortage of oxygen, but dangerous level of carbon monoxide, sulfur dioxide, hydrocarbons and other air pollutants accumulated in several cities. Twenty million people were affected by varying degrees of eye irritation and respiratory diseases. Fortunately, this great blanket of smog was dispersed by upper air currents in the first week of September, and no large-scale health tragedies occurred.

Worldwide water supplies represent a concern for the future of man in terms of both the quality and absolute quantity of fresh water. Many cities have reached or are coming close to their total available fresh water supply and are reaching further and further to obtain new supplies. Los Angeles, for example, is planning tremendous canals and canduits which will bring water from the pacific northwest over 1,000 miles to the growing megalopolis of southern California. Two Russian geographers and hydrologists, Kalinin and Bykov, have pointed out that "The world's water resources from a single entity.

The basic hydrologic cycle of the world depends upon solar energy, atmospheric circulation, heat reflectivity and surface retention. The latter is strongly influenced by the biotic community, kalinin and Bykov point out, on the assumption that the moisture content of the earth remains constant that the total gains and losses from the land and the sea are in balance. Thus, the total precipitation equals the total evaporative loss from land and sea. However, the relative retention time and thus the total amount of water in terrestrial and fresh-water environments can change substantially in relation to the biota and terrestrial storage capacity.

A large proportion of the world's fresh water, falls and resides in relatively unpopulated portions of the globe, particularly Siberia and northern Canada. Here, man's activities are limited by other factors and he cannot readily obtain these vast stores of water. Secondly, the matter of water quality is of immediate concern. The principle of water reuse is now becoming accepted, so that people recognize much of their water has been used previously by other municipal systems. Each reuse cycle however, is increasingly costly in terms of treatment, and each reuse cycle requires disinfectants and other industrial chemicals, combining these problems of distribution and quality, it seems that the water problem of the future is not going to be one of a global water shortage on a planetary basis but one of local crises due to the concentrations of people, industry and water use.

Certainly, the immediate concern is finding local water resources of adequate quality to meet the burgeoning needs of urban populations. To the city dweller in New York or Hong Kong whose water tap runs dry or to the refugee in the Middle East who is dying of thirst, it well be little comfort to realise that 29,000,000 cubic kilometers of fresh water are locked in glaciers and polar ice caps.

Nowhere is scientific controversy more evident than on the issue of world food supplies. Some scientists predict famines and mass starvation within the next 10 years. They feel the stage is irrevocably set for millions of people to die of starvation in the 1980's. In fact, we already have famine in many parts of the world, if one defines famine as people dying of lack of food. Even at the height of the Green Revolution in Asia in which new strains of rice and wheat were producing phenomenally large yields, famine conditions existed in parts of Rajasthan in India: in sections of Burma and in southern Thailand. Starvation is now a grim reality in much of the world and it may well get worse despite the Green revolution. Other scientists refute the pessimism and find evidence for a hopeful outlook in the world food situation. They feel that the Green Revolution which was based on new strains of rice and wheat in Asia and Africa, and which produced a doubling and tripling of yields in one or two years, shows what can be done with modern agriculture. For the first time in many decades, agricultural output increased more rapidly than population growth.

India began to approach self-sufficiency in food production, and the Philippines began to export rice again for the first time since 1900. A word of caution about the Green revolution is also necessary. It is quite true that new strains of agricultural crops can sometimes dramatically increase yields, but often these new strains require more careful management, more intensive care, more water, more fertilizer and more insecticides. They may be more vulnerable to plant viruses or fungi. Although the plant geneticists try to breed in every possible disease resistance, these new strains do not have the centuries of natural selection and regional adaptation behind them as do the local strains which the villages have always used. In West Bengal and Orissa, there were reports that some of the newly introduced high-yield strains of rice were succumbing to a disease that was not affecting that old traditional strains. If this happens to a major extent, the Green Revolution could collapse and a severe blow would be dealt to the villager's faith in the agricultural scientist.

Another factor in the world food picture is one of pure climatic good fortune. Green Revolution in India was accompanied by the best monsoon seasons in over 10 years. A long series of droughts had broken is southern Asia and adequate rains fell in the right pattern. The possibility exists, therefore, that a good part of the success of the Green Revolution was due to good weather, combined with new strains and more fertilizer. This is a question that will be more adequately answered in future year to see if increasing agricultural production can be maintained in years of poor rainfall.

In India too much rain came and it fell to early. The wheat crops in northern India were magnificent in April and May just before harvesting and threshing, but heavy premature rains in many areas caused the wheat to rot before it could be threshed. Bihar had serious crop losses from flooding. Once again, this was a tragic and dramatic demonstration that even the Green Revolution remains dependent on climatic factors outside the control of man. Through all of these arguments on how many people the world can feed, one must ask the realistic question: "Is it necessary that there always be a race between population and food production? Is it desirable to be teetering on the brink of starvation at all times? Is it the best representation of the human condition to be struggling in this manner to feed more and more people at a subsistence level? One also has the impression in reviewing the literature on world food supplies that too much emphasis is placed on future predictions, and not enough on current conditions. We furthermore know that malnutrition exist throughout the world, even in the most advanced countries. We furthermore know that malnutrition of many kind has profound effects upon growth, development and behavior.

Children growing up with malnutrition from early infancy definitely show reduced growth rates and possible retardation of physical and mental development. They furthermore show increased susceptibility to infectious disease. Thus, we return to the issue of the quality of life, and the realization that merely preventing death by starvation does not by any means insure a quality nutritional base for adequate growth and human development. It is apparent that the problem of world food supplies will continue for some time as a problem of both quantity, and quality and especially one of distribution.

In 1958 Paul Sears wrote an important essay entitled, "The Inexorable Problem of Space," in which he concluded that "Our future security may depend less upon priority in exploring outer space that upon our wisdom in managing the space in which we live". The earth's total surface is 196,940,400 square miles of which about 30 per cent or 57,230,000 square miles is land surface. Of this total land surface, approximately 5,000,000 square miles are in Antarctica deserts, and 11,000,000 square miles are in mountainous regions. Of deserts and moutains, at least one-fourth of their jano area cannot be inhabited, so this eliminates another 5,000,000 square miles of land surface for human habitation. We are left with approximately 47,000,000 square miles of land surface, at the most, for the world's population. The world now has nearly about 6 billion people. The world population has doubled in the past 40 years and may double again in the next century, perhaps approaching stability at about 11 billion by the year 2100. By 1990, of the world's 5.3 billion people, 4.1 billion—77 percent lived in the developing world; 1.2 billion inhabited the industrialized countries. India has nearly 1/6th of the world's population. There are 257 person per sq. km. in India. In the Capital City Delhi there are 6318 person per sq. km. Thus it is to say human population is not evenly distributed. It leads to the problem of space.

As with water, air and food, it is important to distinguish between total or potential amount of any given resource on earth and that amount which is actually utilized or practically available to man, it was pointed out in the sections on water and air that the earth's total amount is apparently adequate for many years to come, but the supply in any given city may not be of sufficient quantity or quality to support the standard of living we would like. The amount of living space per person in some cities is less than 20 square feet per person. Theoretically, we could redistribute these crowded people to provide more even spacing around the world. Obviously, this cannot be done, and if it were attempted it might wreak havoc with agricultural production.

The economic and behavioral needs of man are such that he must cluster together, but in so doing, he creates superdensities such as Calcutta, London, New York, Paris, Tokyo, Shanghai and Buenos Aires. The slums of which are definitely deleterious to human health and welfare. Why, we might ask, does man insist on this degree of crowding and then complain about it so bitterly? Why do people continue to pour into the cities of the world when they are already known as centers of population crime and illness?

Perhaps we need to know more about the real spatial and social needs of man. What are his requirements for space and privacy, and how do these interact with his needs for social stimulation and group activity? Does man become conditioned to crowding in early childhood and then seek further crowding throughout his life? Does he require the social pressure of a crowd because he feels a compulsion to be in touch with the group and thus be able to predict what the group is going to do next? Have we produced, in our urban centers, self-destructive, systems which so modify the behavior of their inhabitants that these inhabitants exaggerate their own problems and can no longer achieve rational solutions? These are all tough question that cannot be answered in simple or easy terms. The answers will require the best intelligence we can muster from many disciplines. Some sorts of answers will certainly be necessary before we can even begin to solve the ecologic and behavioral problems associated with our own use of terrestrial space.

With all of the last four life-support requirements of man-air, water, food and space—we have seen that the major problem of immediate importance lie in distribution more than in total amount. Technologist therefore look upon these problems as soluble by technological means. They feel, they are essentially matters of economics and transportation Behavioral and environmental scientists see them as more basic, however, relating to man's fundamental ecologic behavior. They feel, they are essentially matters of economics and transportation.

Behavioral and environmental scientists see them as more basic, however, relating to man's fundamental ecologic behavior. They feel that certain basic social and ecologic tendencies of man prevent equitable distributions of the earth's resources. Obviously this type of argument can go on forever, and it is perhaps necessary to take a pragmatic approach and look at the current picture for the best glimpse of the future. Many of the industrial nations of the world, especially the United States, the national northern Europe and Japan, have enjoyed rising prosperity for the last 25 years. In ecologic time, of course, this is but a flash in the pan, but let us assume that it can be sustained for another 25 or 50 years. It is worth examining where the material and energetic support for this prosperity arises. This distributional inequality is another major symptom of ecologic instability in the world. One would hope for some narrowing in the gap between the haves and the have-nots, but unfortunately the opposite trend is still occurring.

The wealthy nations of the world are far outstripping the poor nations in their rates of economic gain and human betterment. There are, however, self-corrective mechanisms within this system as, for example, when some segment of the labour it can find which will do a satisfactory job. Thus, there has been a great expansion of manufacturing in Japan, Hong Kong, Taiwan and Singapore in industries as diverse as steel, electronics, textiles, toys and optics.

The Western factory worker experiences for the first time an improving situation. The western industrial worker continues to refuse his own working prospects by striking for higher and higher wages. This will certainly be one mechanism by which wealth is more adequately balanced between nations. Still the mass of poverty throughout much of Asia, Africa and Latin America is so great that it is difficult to imagine a painless equilibrium. The differential distribution of wealth remains sharp and represents a type of ecologic situation which cannot last indefinitely without severe consequences and reverberations.

To the ecologist it means either a successional stage which is not yet mature, or a system headed for a major breakdown. Economic and ecologic maturity should ideally be in the direction of broader and more equitable distribution of the earth's resources—a narrowing of the gap between the haves and the have-nots, both within and between nations, provided the total system does not pass beyond the state of corrective change.

Any ecological system or biotic community can be overloaded to the point of collapse by a variety of factors. Certainly one of the factors for man of most immediate and realistic danger is that of social violence. Either by war or by civil riot, man threatens his own existence. To maintain a sense of physical security man has built an incredible stockpile of destructive power.

The nations of the world spend more than 200 billion dollars per year on armaments—more than on health education, welfare, agriculture, child care and care for the aged combined. Twenty tons of explosives are stored in the world for every man, women and child. Russia and the United States have stockpiled 6,000 nuclear war heads possessing thousands of megatons of explosive force. Each megaton has more destructive power than which existed in any of the previous wars of history.

The world spends almost 8,000 dollars per year training each soldier and less than 100 dollars per year educating each child. The question is, of course, how long can man continue to preside over armament like this without using it? And the answer seems to be, not very long. The fact is we do use part of the power in an attempt to achieve social goals. It is being used daily in various parts of the world to maintain territory and dominance ranks. It is justified as a spare and deliberate use in hopes of avoiding the all out effort.

How long this can continue is anyone's guess, and the ecologist can claim no more omniscience than the political scientist. The ecologist points out, however, that the ecologic and social pressures which generate international military tension may rise with increased population pressure. It is not likely that man can solve the problems of war and violence until he approaches and understands their basic orgins in terms of the behavioral and ecologic relationship between social groups, and the environment.

Many scholars feel that one of the great problems of the present and the future is the rapid rate of change which is occurring in society and technology. There is not, however, agreement on the consequences of the rapid change, changing life patterns are too rapid for man to adapt successfully. Mankind is so overstimulated by the increase in new sight and sounds, so bombarded by world wide mass media, and so constantly beset with noval situations and new problems that he will break down psychologically and physically. His responses will be maladaptive and pathologic. He will thus collapse under sensory overload.

It is quite possible that each of these points of view has large elements of truth in it. In the present world one can find many examples of both. Our mental institutions are already filled with those who could not adapt to sensory overload and who experience a personal breakdown of physical and mental health. Yet one can find people are living a sub-human existence where the quality of life is so low that the amount of human pain and misery seem unbearable. Yet man has adapted successfully to these inhuman conditions. In many ways, the future is with us now. We can find throughout the world an abundance of high quality life in high quality environments, but we can also find an abundance of low quality life in impoverished environments.

Many of the questions about the future revolve, therefore, around the question of relative balance and proportions. Which way are we heading and which way will the scales tip? A major aspect of the future centers around our sense of responsibility to posterity. No representative of posterity is here to speak for the segment of society. How do the present populations of mankind regard their obligations to future generations? Are we willing to be more conservative in our life requirements, and are we, in fact, willing to reduce our rates of personal economic growth to insure a better environment and a greater abundance of resources for future generations? This may be one of the central issues of the next twenty years, and one on which the quality of present and future environments depends.

Environmental problems have become a global issue: Some of the world wide issues are briefly described below:

(a) Climatic changes/Green house effect

Many studies show that the atmospheric pollution is on a global scale. Today, however, it is not only the air over our cities that is polluted, the entire atmosphere of our planet is now afflicted to some degree. Meteorologists talk about a nebulous veil of air pollution encircling the entire earth. Smong has been observed over the North Pole, and in their unlikely places. There are range of environmental pollution issues originating largely in developed countries which are also of global significance. Carbon dioxide (CO_2) build-up leading to climate change is one of those key issues. Recent research indicates that increasing concentrations of CO_2 in the atmosphere, due mainly to the burning of fossil fuels could cause a warming trend leading to climate change in the next century. While there is still scientific uncertainty about the pace and impact of such a change it could be of sufficient magnitude to produce major physical, economic and social dislocations on a global scale. A special challenge posed by the issue is that if preventive actions are postponed until some effects are perceived, it may be too late to avoid appreciable climate change. It is therefore in the interests of all nations to support international efforts to understand the phenomenon at early enough stage for action to be taken to reduce or avoid major economic, social and environmental disruptions around the world.

However, as developed countries are currently responsible for over two thirds of total world emissions of CO_2 by fossil fuel combustion, they have a special responsibility for supporting and contributing to accelerated efforts to assess more precisely the nature of the problems and the alternatives for action to reduce or avoid any serious risks. Another issue is the world-wide distribution and use of harmful or potentially harmful chemicals. There are now some 70,000 chemicals on the commercial market, and many are currently used and released into the environment with little or no knowledge of their potential long-term effects. It is estimated that another 1,000 new chemicals enter the market each year. Chemicals cross national boundaries in a variety of ways: they are transported by winds and water.

(b) Risks to Ozone Layer

Ozone is a different form of life sustaining oxygen, having three atoms instead of the usual two. But this third atom, scientists have found, is very unstable and can be eliminated by chlorine ions. The scientist have now found a close link between the family foam mattress, refrigerator and the air conditioning system of an average home and Ozone layer, 50 to 70 kilometers above the earth. They believe that free chlorine ions into the atmosphere, from plastics and other chemicals remove that extra atom of oxygen which makes Ozone.

Only, in this case, the loss of the one atom changes the ability of the layer to act as a strainer to absorb the damaging ultraviolet B wavelengths from the sunlight. The findings by the experts have revealed that the rate of Ozone Layer depletion is sufficiently large to present serious consequences for man, animals and plants. In addition, new threats to the ozone layer are being recognized. Take, for example, the refrigeration industry. Food preservation has become a necessity and refrigeration is here to stay. Only advanced technology can provide the solution to prevent the release of the damaging ions.

Another example is nitrogenous fertilizer which, it is suspected, release chemicals damaging to the Ozone layer. The demand for more food and the booming green revolution in many developing countries have pushed up the use of this fertilizer. Today the world uses about 50 megatons of it. In another 20 years the rate of uses will be 100-200 megatons. Though the Ozone Layer accounts for less than one millionth part of the earth's atmosphere, it is of vital importance to the earth's life form. It has two main functions: first it absorbs the UV-B while allowing the heat generating infra-red rays to reach the earth freely. The heat from infrared rays sustains plants and human lives by keeping the temperature of the planet at a desired level. Secondly, it influences the whether by stabilizing the stratosphere which acts as a cap to the turbulent weather system in the troposphere.

Scientists knowledge of the Ozone layer is still limited for want of advanced technology. The presence of other substances in the stratosphere, which absorbs UV-B, makes it difficult to come to a numerical and logical conclusion. But the scientists do agree that an average 70 per cent of the total UV-B reaching this planet falls on the equatorial regions between the latitudes 30° S and 30° N. They also agree that for genetic reasons people with light eyes and light skin (including albinos) are most affected by UV-B. Scientists today claim UV-B can destroy micro-organisms outright as well as cells in plants and animals. It also destroys the DN molecules and the genetic cell composition. Under natural conditions plants and animals have developed behavioural adaptation but gradual or sudden increase in UV radiation due to changes in the Ozone Layer is likely to disturb this balance. While the scientists agree that UV-B contributes to the induction of cancer they agree the dose response relationships are too complex and unknown. These make the quantitative analysis speculative, as regards the cause of melanoma cancer and resulting deaths.

The North European Conference of Scientists November, 1980, warned that non-melanoma skin cancer in light-skinned people would increase and many crops and many acquatic organisms, including shrimps ad fish eggs, could be damaged by increased UV-B infiltration. This could even affect the reproductive process of plants and animals. It recommended that existing satellite and ground based ozone observing system should be integrated through Global Ozone Observing System (GOOS). They also recommended updating of equipment and setting up of more stations for observing the ozone system. The scientists concerned with outer limits of the biosphere felt they are not yet technically equipped to measure accurately the rate of ozone depletion-already depleted by one percent. Though the danger is known, science has not yet provided a perfect model of measuring the depletion rate.

(c) Acid Rain

Every one realises how important the rain is. Without it the continents would be barren. Yet now the rain in parts of the earth has taken on a new and threatening complexity. It mixes in the air with pollution from burning fossil fuels particularly in power stations, factories and motor vehicles and brings down dilute sulphuric acid. This is killing fish and water life, and corroding buildings, including some of the world's most important ancient monuments. It may also damage forest and croplands, and possibly pose a substantial threat to health. Acid rain is now recognized as an international problem. The air of industrial towns partly is cleared by chimneys at factories which push the pollution high into the air. These chimneys have made the things better locally by dispersing the pollutants, but aggravated the international difficulties. For the sulphur and nitrogen compounds emitted by burning fossil fuels can be blown thousands of kilometers by the winds, to cause acid rain in countries far from their points of origin.

A special conference on acidification of the environment was held in Stockholm in 1982. It reviewed and assessed a large amount of scientific information not previously available. So far, acid rain has been seen as an essentially regional problem, confined to the industrial areas of the northern hemisphere. But, though the problem was first perceived there, it may become much more widespread for acid rain is likely to occur wherever fossil fuels are intensively used. Acidification is an environmental problem, or becoming one, in parts of Europe and North America.

Similarly polluted areas are likely to exist elsewhere in the world, especially around large urban and industrial coagulumerations. Soils are normally much better able to resist acidification than lakes, rivers and streams. Meanwhile, acid accelerates corrosion in most materials used in construction of historic buildings, and other important cultural objects. Some of th worlds greatest cultural treasures including Parthenon in Athens and Trojans' column in Rome are being eaten away by acid fall out.

There must be more effort to research and develop ways of improving energy conservation, environmentally appropriate technologies for producing heat and power, and techniques for removing sulphur from fossil fuels and gaseous emissions.

The problem of acidification may well spread to new areas as a result of rapid industrialization and the growth of cities in other parts of the world, particularly developing countries. So it is important that areas affected by acid deposition and susceptible to damage from acidification are identified as soon as possible. It they are, the damage could be mitigated or even avoided of a minimum cost to society by initiating research and applying what we already know about pollution control.

(d) Pollution of World's Oceans

The data collected from the Baltic, the North Sea, and some North American and Australian stuaries and coastal waters suggest that in most of these areas the pollution that caused most concern was due to sewage, agricultural chemicals, oil and metals. Metal concentrations were clearly elevated in coastal waters, and in fish and shelfish living there. In some areas, mercury levels in species such as tuna were high enough to make these fish unsuitable as human food.

Some of the pollution came via river; the amount of iron, manganese, copper, zinc, lead, tin and antimony that reached the sea by this route was far greater than would be supplied by natural geological processes. Other contaminants came through atmospheric deposition: the importance of this pathway for metals and synthetic chemicals was increasingly recognized during the decade. Offshore oil and gas exploration and dredging for sand and gravel in coastal areas increased during the period 1972-1982.

Coastal zone development affected extensive estuarine areas, as well as mangrove swamps and coral reefs. Oil pollution killed sea-birds fouled beaches and affected tourism. Although tanker accidents were the source of less than 5 percent of all the oil entering the sea, accidents released large volumes in small areas, and were therefore especially damaging.

A recent report issued by a group of experts on marine pollution and published by the United Nations Environmental Programme, Warns that increasing contamination of the coastal seas, which has so far only caused localized disturbances to the marine environment may become much more serious.

The combined effects of the dumping of sewage sludge industrial wastes, litter and oil spills into the Mediterranean, the North Sea, the Baltic and the Gulf of Mexico, may well cause ecological disasters which could gradually lead to a global crisis since the oceans are a unified life support system. The report points out that the sea's capacity to absorb wastes has its limits, so that as the world economy continues to grow, even more sewage, industrial wastes and other effluents are likely to find their way into the sea.

Man-made radioactive materials have been discovered in the open ocean and scientists have found traces of DDT in th Arctic sea and the Antarctic penguin, thousands of miles away from the areas where the pesticide is used. And because many of the disturbances which are introduced occur at vulnerable points such as the surface of the sea or on the sea floor, where oceanic life-supporting processes are generated, the dangers and magnified. Although the modification of this microlayer has occurred only on the vicinity of oil spills and around the marine dump site of municipal wastes and no evidence has yet been found that detergents in the open ocean have significantly influenced the properties of the sea surface, the report points out that environmental catastrophies in certain coastal areas and seas might already have occurred. Only improved management of the coastal areas and seas which involves controlling more closely the dumping of pollutants and careful monitoring of contaminant levels, will prevent a further deterioration of coastal environments. Much work however is still to be done before the several ecological sub-systems which constitute the ocean are fully understood. By the end of 1982 action plans were in operation or planned for 10 regions involving some 100 coastal States. Coastal and marine national parks and reserves were extended.

(e) Desertification

Desertification is now perceived as an intensifing world wide threat which, to varying degress, affects over 100 Member States of the United Nations. More than 20 percent of the world's population (850 million people) live in the 35 percent of the Earth's land surface (45 million km) which consists of the arid, semi-arid and sub-humid zones which is at risk of desertification. Three quarters of this area is already at least moderately desertified. Each year an additional 21 million hectares of agricultural land deteriorates through desertification to a point at which it is no longer economically productive. The last production has been valued at 26 billion dollars annually.

Although desertfication is exacerbate by severe drought, its principal cause is human over exploitation of dry lands through over-cultivation, overgrazing, poor irrigation practices and deforestation. When drought strikes the overtaxed livelihood systems, desertification results. The most devastating environmental and economic effects of desertification are in developing countries in the drylands.

National balance of payments problems and local rural poverty lead to over-exploitation of lands and forests for food, fuel, cash crops and meat for export. As these and other pressures increase to unsustainable levels, agricultural productivity and soil fertility decline, creating in turn new pressures which lead to further degradation. Recovery of land for agricultural purposes become costly and often economically unfeasible, thereby reducing the economic potential of the affected countries. In many of these countries agricultural land is also being reduced by salinization, urban sprawl and sand-dune encroachment. Increased desertification in one part of the world can indirectly induce desertification else where, as loss of agricultural land and increased food demand raise Prices and intensify pressures on marginal drylands in other food exporting and importing countries alike.

There are also other impacts on international trade, to the extent that developing countries affected by rapid desertification will have a reduced capacity to import goods and services from developed countries and from other developing countries. Other globally shared impacts include loss of valuable genetic resources, disruption of hydrological cycles and increase of atmospheric dust. A great deal is already known about the causes and effects of desertification and about the measures needed to halt or combat it. Moreover, there is favourable cost benefit ration to justify, preventive action. The 1980 estimate for implementing the 1977 plan of Action to combat Desertification was 2.4 billion dollar annually over 20 years, which is only a fraction of the estimated value of agricultural production lost each year through desertification. It is, therefore, in the interests of all countries to establish some concreate cooperative actions between groups of developed and developing countries to halt further desertification, a goal which can be achieved, given a sufficient allocation of resources, on the basis of currently available knowledge.

(f) Loss of Tropical Forests

Although tropical forests are found in early half of all United Nations Member States, most are in twenty countries and nearly half their total area in three countries alone (Brazil, Indonesia and Zaire). The tropical forests are the world's richest biological zones, and are estimated to contain as much as 40 per cent of all terrestrial species on the planet. In addition, tropical forests provide a wide range of useful products of economic significance for both developing and developed countries. Undisturbed tropical forests are also home to millions of the world's tribal peoples. Many experts contend that tropical forests are being exploited at a rate and in a way that is ecologically destructive and economically unsustainable. There are varying estimates of how much of the tropical forests will be lost by the year 2000 through complete conversion to other uses and by severe degradation.

The results of the tropical forests resources assessment project conducted by FAO and UNEP indicated an estimated minimum loss of closed tropical forests of 12.5 per cent by the year 2000 (an average of 7.5 million hectares annually between 1980-2000) through complete conversion to other uses alone. Other studies which not only consider complete conversion but also include severe degradation caused, for example, by logging operations, by over-grazing, by over-exploitation for fuelwood, or by natural causes such as fires, insects and diseases, have estimated losses of tropical forests as upto 20 million hectares or more in a year.

The FAO/UNEP study estimates that tree planting on industrial and non-industrial plantations accounts for only 10 per cent of the area of tropical forests cleared annually. The major causes of deforestation differ by region. According to the FAO/UNEP report in Latin America the two major causes are shifting cultivation and the conversion to pasture land for cattle ranching. In Asia and Africa, shifting cultivation by subsistence farmers and removals for fuelwood are major causes.

The extraction of timber from tropical forests for industrial purposes is not considered by the FAO/UNEP reports as a major factor in deforestation. On a world-wide basis, the UNEP report estimates that fuelwood and charcoal removals from tropical forests are as much as eight times greater than for industrial wood. Serious degradation of loss of watershed forests in particular has a wide range of major ecological and economic effects through increased erosion, floods, landslides, and silting of hydro-electric facilities, irrigation systems reservoirs and harbours. The lives and livelihood of half the world's population depend directly on the wise management of watershed forests and eco-systems.

Deforestation also influences local, regional and possibly even global climatic conditions by altering hydrological regions, reducing vegetation cover and causing more heat and light to be reflected back to the atmosphere (the albedo effect), and through accelerated burning of the forest biomass, which contributes to carbon dioxide build-up in the atmosphere and the global warming trend. The economic, social and political implications of possible regional and global climate changes have only begun to be assessed.

(g) Soil Loss

Almost every environmental concern connects clearly with the importance of soil. Soil acts as a reservoir for water and other inorganic substances cycling through the eco-system. It provides habitat for many of the different populations or decomposer organisms that control the rate and capacity of nutrient recycling. The importance of soil to agriculture can hardly be over emphasised. Agriculture provides about 44 percent of the poorest countries gross domestic product and 83 percent of their employment. Growing more food, building irrigation projects, conserving nature and precious wildlife and other needs must depend on proper soil management. The wide-ranging importance of soil confirms that soil management must consequently be many-sided. The broad range of natural and man-made stress that causes the two type of soil damage: physical displacement and alterations in soild structure and composition.

Uncontrolled shifting cultivation, salinization and water-logging flooding and plant cover destruction are same. An appropriate use of chemicals, weapons testing and deployment in warfare are some other casuses of soil nutrients and other types of soil degradation. Soil damage is also caused by patterns of dry land mountain side cultivation that carried out without arrangements for bunding and terracing to protect the topsoil. Continuous cropping to meet growing food needs leave title time for soil conservation techniques. Deterioration of the productive capacity of the soil is most acute in areas affected by erosion, drought, mineral stress and other related factors. Soil resources that are vulnerable to these factors constitute .82 percent lands in South Asia.

While erosion is an integral part of the natural system, occurring even when land is covered by grass, or trees, its extent and intensity can be minimized provided land denudation is correspondingly reduced. The fact that it takes nature 100 years or more to build one inch of top soil should have engineered the vigilant observance of soil conservation measures. Instead, problems of soil erosion have continued ticking. Countries in East and South Asia are particularly burdened with soil erosion occuring in watersheds. Anticipated acceleration of agricultural production and other activities between now and 2000 is casting its shadows: existing cultivable land will be further subjected to intensified use as cultivation of unsuitable areas persists; logging, firewood gathering and grazing will accelerate the elimination of vegetative cover, while the construction of roads, mine and water resources projects will be further pursued in the course of development. Such prospects place the least developed countries in an even more precarious situation. Slash and burn farming is a common enough culprit. Wielding destruction in gradual succession, it initially causes the loss of fertility by the thin topsoil, and later eroding its stability. In India, slash and burn farming is particularly rampant in the north-eastern and central regions affecting between six and eight million hectares of land.

Salinization affects 80 million hectares (8 percent) of total land in South Asia. Something should also be said about agricultural techniques which are designed to raise land productivity but instead lead to soil loss. The use of cheap nitrogen, for example, replaces nitrogens, fixing legumes, resulting in loss to topsoil. The commonality of problems, in addition to the implications transboundary soil management points up to the fact that soil problems cannot always be resolved within the context or national concerns. Clearly, what are required are mutual policies and strategies concerning land management, and the World Soils Policy provides an appropriate starting point.

Working in collaboration with the relevant UN bodies, UNEP has proposed the World soils Policy programme which endeavours to enhance international awareness, assist countries in the formulation of national soil policies, help develop technical and scientific knowledge and collect, compile and disseminate data on the use and management of the worlds soil resources.

(h) Depletion of genetic resources

25,000 species and more than 1,000 vertebrate species and sub-species are now estimated to be threatened with extinction, yet these two categories of organisms represent only a small proportion of the terrestrial species on the planet. According to some estimates as much as from half a million to a million species would be extinguished over the next two decades. Extinction of species on this compressed scale would be without precedent in human history. As much as two-thirds of all terrestrial species and the majority of endoangered species are located in developing countries. A significant loss of species in developing countries would have a world wide effect on future agriculture, industry and science, and on human health and welfare generally. The most serious threat to many species is the disruption and destruction of their habitat by man, especially in the tropical forests which alone harbour as much as 40 per cent of all terrestrial species.

Significant species loss could directly affect human health and welfare locally, and even nationally, in many developing countries, where wild plants and animals are a source of fodder, traditional medicines and income especially for the poor in rural areas. In some countries tourism largely on wildlife is a major source of foreign exchange. In many developed countries over one-thrid of the gain in the productivity of many major agricultural crops over the last few decades has been attributed to breeding programmes using genetic material from related wild plant varieties, many of which are found in developing countries.

Moreover, many of the major crops in developed countries now have a limited genetic base. For example, more than half the Canadian prairie wheatlands are devoted to a single variety, Neepawa, which was developed with germplasm from Kenya; 72 per cent of United states potato production depends on only four varieties; and the entire United States soyabean industry is derived from a mere six plants from Asia. A significant loss of relavant wild species could limit the possibilities for maintaining or increasing the yields of major crops, such a wheat, maize, rice potatoes and for combating new strains of diseases or pests which threaten them. Genetic stocks from one part of the world frequently make an enormous contribution to another. Current research shows that enormous opportunities lie ahead, such as the utilization of Asian blue-green algae of fix nitrogen for crop fertilization, which may constitute a gain of hundreds of millions of dolars a year to the economics of developed countries alone.

Major medicines and other pharmaceutical products in developed countries are also often dependent on plant and animal species. For the United States of America it has been estimated that more than 40 per cent of the prescriptions written each year contain a drug of natural origin-from higher plants (25 per cent), microbes (13 per cent) or animals (3 per cent) as the sole active ingredient or as one of the main ones.

It has been estimated that among the 10 percent of species that might be extinguished by the year 2000 there could be many plants with potential for use in new plantbased drugs and pharmaceuticals with a combined world wide commercial sales value of as much as 200 million pounds annually. Biomedical research in developed countries is also vitally dependent on wild species, many of which are threatend with extinction.

The innovative capacity and future of the rapidly expanding field of bio-technology, practised largely in developed countries, relies considerably on genetic patterns and resources from many wild plants and animals as only about 10 per cent of the world's plant species and 1 per cent of the animal, species have so far been subjected to even a preliminary screening for potential medical or commercial applications, a significant loss of species could also represent a permanent loss of sources for more effective chemicals, medicines and other useful products. In spite of this interdependent relationship, species loss continues to accelerate.

Developing countries often lack the capacity and economic incentive for extensive species protection. They have derived little or no economic benefit from the development and use of genetic materials in developed countries, where the scientific and technical expertise to exploit species and their genetic resources predominantly resides. Nevertheless, developing countries are and will remain the principal source of genetic materials from wild plants and animals. This clearly shows that there is a common incentive for developed and developing nations to prevent a potentially catastrophic loss of species. The mutual advantage of comprehensive joint effort could result in new medicines and other useful products that would contribute to improved levels of human health and welfare in all countries.

Human progress has always depended on our technical ingenuity and a capacity for cooperative action. These qualities have often been used constructively to achieve development and environmental progress: in air and water pollution control, for example, and in increasing the efficiency of material and energy use. Many countries have increased food production and reduced population growth rates.

Some technological advances, particularly in medicine, have been widely shared. But this is not enough. Failures to manage the environment and to sustain development threaten to overwhelm all countries. Environment and development are not separate challenges: they are inexorably linked. Development cannot subsist upon a deteriorating environmental resource base: the environment cannot be protected when growth leaves out of account the costs of environmental destruction. These problems cannot be treated separately by fragmented institutions and policies. They are linked in a complex system of cause and effect.

First, environmental stresses are linked one to another. For example, deforestation by increasing run off, accelerates soil erosion and siltation of rivers and lakes. Air pollution and acidification play their part in killing forests and lakes. Such links mean that several different problems must be tackled simultaneously. And success in one area, such as forest protection, can improve chances of success in another ares, such as soil conservation.

Second, environmental stresses and patterns of economic development are linked one to another. Thus agricultural policies may lie at the root of land, water, and forest degradation. Energy policies are associated with the global greenhouse effect, with acidification, and with deforestation for fuelwood in many developing nations.

At these stresses all threaten economic development. Thus economics and ecology must be completely integrated in decision-making and law making processes not just to protect the environment, but also to protect and promote development. Economy is not just about the production of wealth, and ecology is not just about the protection of nature: they are both equally relevant for improving the lot of humankind.

Third, environmental and economic problems are linked to many social and political factors. For example, the rapid population growth that has so profound an impact on the environment and on development in may regions is driven partly by such factors as the status of women in society and other cultural values.

7 Ecology of Populations

A population is a group of living individuals set in a frame that is limited and defined in respect to both time and space. Also population is an interacting, or potentially interacting group of individuals usually of the same species. It may be the population of ants in an ant hill, the population of salmon in a river, the population of a deer in a state and so forth.

Biologically, a population should have some natural boundaries as is the ants in an ant hill, but the term is often used with artificial boundaries in mind, as in referring to the population of deer in a state. A state might have several biological population of deer, but for economic or political consideration it is convenient to put them together. The human population representing varied ethnic religious or economic groups these population within one nation might have different characteristics.

In the initial study of populaion it is helpful to recognize certain attribute of population which distinguish them from individual organisms like organisms have a structural organization, a functional unity, and a pattern of growth and development unlike individuals they have group attributes and statistical properties that no single individual possess. The more important of these population characteristics or group attributes are as follows,

— *Density:* Population size in relation to unit of space.

Birth rate: or more broadly natality the rate at which new individuals are added to the population by reproduction.

— *Death rate:* or mortality rate: The rate at which individuals are lost by death.

— *Dispersal:* The rate at which individuals immigrate into the population and emigrate out of the population.

— *Population growth rate:* or growth form. The net result of natality mortality & dispersal.

— *Dispersion:* The way in which individuals are distributed in space, generally in one or more the following patterns:

i) Random distribution, in which probability of an individual occurring in any one part is the same as the probability of it occuring at any other part.

ii) Union distribution, in which components occur more regularly than random such as corn in a cornfield.

iii) Clumped distribution (the most common in nature) in which individuals or other components are more irregular than random as for example a clump of plants arising from vegetative reproduction, a flock of bird, of people in a city.

— *Age distribution:* The proportion of individuals of different ages in the group.

Genetic characteristics especially applicable to population ecology as for example, adaptiveness reproductive fitness and persistence. Individual organisms are born, have age, grow and die but such characteristic as birth rate, death rate, density and other listed above are meaningful only at the group level. If we are to understand thoroughly the ecology of a species, we must study and measure these population group characteristics as well as know the life history and identifying features of the species. And of course, we all must be concerned about the vital statistics of human population.

Some of the most important aspects of population structure are population size, density, spatial distribution age and sex rates, breeding structure and social organization. Population sizes may vary from a few individuals to millions of individuals. The population of whooping cranes (Grus American) numbers about 50 individuals a single group of birds barely clining to existence. Other rare and endangered species around the world also represent very small populations.

The world's population of the Indian lion (Panthera Ieo) numbers less than 200 individuals living in the Gir forest of western India. The California condor (Gymonogyps californicus), golden marmoset (Leontideus rosalia) Javan rhinoceros (Didermocerus sumatrensis) orangutan (Pongo pygmaeus), Gray jungle fowl (Gallus sonnerati) and many other species are now represented by small remnant populations wavering on the edge between life and death. At the other extreme, the world contains many species of tremendous abundance.

A population of starlings in a sigle winter flocking roost in eastern United States may number 5 million birds. The white-tailed deer population of Pennsylvania was estimated in 1946 to contain 1 million deer; the rhesus monkey population of one province in northern India was estimated in 1960 to contain between 800,000 and 1 million monkeys (it now probably contains less than half that number); the Norway rat population of

Baltimore was estimated in World War II at 400,000; and the bison population of the Great Plains was estimated to contain 60 million individuals in the early nineteenth century, though by the 1890's the bison population had crashed to less than 600 through the greed and disruptive influence of man.

Sensusing wild animal population is fraught with many difficulties of sampling error and statistical bias. Occasionally direct counts are possible as in aerial photographs of big game herds or sea bird colonies, but for most animals this is impossible and various kinds of sampling methods must be employed. Most animals are not readily visible because of their behavior and habitat, or because they exist in such abundance or scarcity that they cannot be readily counted. It therefore becomes necessary to estimate numbers through programs of capture and resembling.

With many animals systematic program of capture-mark-recapture has been helpful in obtaining estimates of abundance. The ratio of marked to unmarked animals in subsequent trapping runs provides a population estimate known as the Lincoln Index. If, for example one is studying a resident bird or rodent population in a forest, and succeeds in capturing and banading 100 individuals all of which are released, and then if on a second trapping program, one again catches 100 animals of which 20 were previously banded one would estimate the total original population at 500 animals. This is based on the Lincoln Index ratio which may be written.

$$\frac{P_1}{M_1} = \frac{T_2}{M_2}$$

Where P equals the unknown population, M_1 equals the total number of individuals marked in the first capture period, T_2 equals the total number of individuals captured in the second capture period, and M_2 equals the number of those in the second capture period which were marked.

The validity of the method involves several major assumptions: (1) that the marked animals mix randomly in the population, and (2) that the probability of recapture is the same for each individual regardless of whether or not it is marked, and (3) that there be no immigration or emigration;, death or births between the sampling periods. Actually, these conditions are rarely attained in natural populations.

Many studies on birds and mammals have shown that some individuals are easily captured whereas others are rarely captured. There are thus "trap-happy" individuals and "trap shy" individuals. In some species, one experience with trapping on the part of an individual animal may reduce the probability of subsequent recapture, whereas in others it may enhance it. Such behavior patterns bias the data and reduce the value of the population estimates made with this method. Also, birth, death, immigration and emigration are often continuous processes within populations, so their effect can rarely be eliminated entirely.

Nonetheless, the Lincoln Index approaches validity in certain cases. It can be a helpful tool for population estimates, particularly in the following cases: (1) if a capture method is available which is relatively nonselective and random in its action, (2) if a relatively confined population is available so that emigration and immigration do not occur, as an island population, a woodland population surrounded by forest, an oasis , etc., (3) if the population can be sampled at a time when it is relatively stable, not reproducing during the sampling period, and not migrating.

Another method of estimating animal population size is based on a change in sex ration before and after a known harvest of one sex has been removed. This has been called the Kelker Ratio, and has been used in the estimation of deer, pheasants, and other wildlife populations where hunting statistics are available. In a study of California deer by Dasmann, the sex ratio observed before the hunting season was 74 bucks to 139 does, and after the hunting season in the same area it was 50 bucks to 139 does. The harvest in the hunting season was known to be 246 bucks. Hence the removal of 246 bucks population should be in the same proportion to the original sex ratio as the total buck harvest was to the change in sex ratios produced by the harvest. Thus, the following delker formula was applied to calculate to total population of bucks.

$$\frac{S_1}{P_1} = \frac{S_1 - S_2}{H}$$

Where S_1 equals the pre=hunting sex ratio, P_1 equals the pre-hunting buck population, S_2 equals the post-hunting sex ratio (therefore S_1—S_2 equals the change in sex ratio during the hunt), and H equals the total harvest. One can readily see some of the observational problems and potential biases in the Kelker Ratio as the Lincoln Index. The Kelker Ratio assumes that the observations of sex ratios before and after harvests are equivalent.

It the buck become shyer than the does and are harder to observe after the hunt, this enters a major bias. It is essential that the field ecologist be fully experienced and knowledgeable in the habits of the animal he is studying before he applies these population estimation techniques. He can then recognize the sampling problems and possible sources of error. If reasonably unbiased sample estimates are obtained, it is then important to subject them to statistical treatments such as standard error calculations, or analyses of variance to evaluate the sampling errors which might occur by chance alone. It is thus important for ecologist and statisticians to work closely together in population studies.

Population density is an important aspect of both population size and distribution. In many populations, the precise boundaries of the population are unknown. Though the distributional limits of the species may be recognized each species usually consists of several populations. The essential feature which distinguishes a population is whether or not the individuals are potentially interactive, or potentially interbreeding.

If a species consists of two or more discrete aggregations, which do not interact or influence each other these aggregations can then be considered populations. If this condition persists over an extended period of time, different races, subspecies and even species may emerge as gene frequencies change and evolution occurs. Since populational boundaries are often unknown the most practical description of population abundance is frequently in terms of densities.

As an example many ecologic studies of small rodents have shown population densities usually in the range of 0.5 to 20 animals per acre. But showed population densities of the deer mouse (Peromyscus leucopus)in southern Michigan woodlots to range from 3.08 to 10.32 individuals per acre, Blair estimated the population density of beach mice or old field mice (Peromyscus polionotus) to range from 0.83 to 1.41 individuals per acre. Ashby estimated population densities of the bank vote (Clethrionomy glaeolus) a relative of the meadow mouse, to run as high as 17 to 18 individuals per acre in central England.

An important consideration, which will be emphasized later, is that many populations fluctuate widely in density and size in the course of natural events. Occasionally small mammal populations reach amazingly high densities. In California, Krebs described a meadow mouse or vole population (Microtus Californicus) which achieved a density of 150 to 300 individuals per acre in a relatively small area, and Parson described a population of house mice in California which increased from a density of less than 5 mice per acre to 300 mice per acre within 6 months.

At these high levels, or course, rodents car become agricultural problem and can precipitate economic losses of crops or stored food products. Splitte has estimated that wild rats (Bandicota bengalensis) in Calcutta grain storage warehouses may achieve densities of one rat per sequare meter, equivalent to 4,000 per acre. These represent extreme concentrations, or course, but they indicate the population potential of some species.

Another important consideration in concepts of both population size and density is the spatial distribution of individuals. Spatial distribution involves a fascinating interplay of behavioral and ecologic factors, and is a prime example of the inseparable tie between behavior and ecology. We have been speaking of population densities as though animals were randomly distributed. This is rarely true, and it is much more common for population to be clumped and nonrandomly distributed.

The reason for such clumping or nonrandom distribution might be numerous. In some cases clumping would be attributable to the heterogeneity of the environment the distribution of food, cover and shelter. For example, dung beetles in an old field would be clumped around the dung isopods and millipedes around the fallen log, etc. In other cases, the clumping might be due to the social behavior. This clumping may be related to environmental resources, but not necessarily so. This is not to say that random dispersion of animals never occurs. Park pointed out that flour beetles (Tribalism confusum) are distributed in flour according to a Poison distribution, indicating that the individuals are spaced at random.

Cole found that certain spiders were distributed randomly under boards placed on the forest floor. These examples are probably exceptions to the general rule, and they are most likely attributable to either the homogeneous or artificial nature of the environment in each case. For purposes of generalization it can be stated with reasonable assurance that most animals and many plants in natural environments show a clumped or nonrandom pattern of distribution.

It is obvious that animals within a population differ in a number of ways: sex, age, breeding condition health, physical condition and social status. Demography, or the numerical analysis of populations, is concerned with the sex, age and breeding condition of the population. Medical science and epistemology are primarily concerned with the health and physical condition of the population and behavioural science is primarily concerned with the social and behavioral states within the population. In most vertebrate populations primary sex ratios at hatching or birth approach 50 percent male and 50 percent female. In some forms, there is a slight deviation from this; in man, for example sex ratio at birth is closer to 52 percent male, and this is true for rabbits cattle and many birds.

In domestic chickens, sheep and horses, sex ratios at birth or hatching are usually closer to 49 percent male. The secondary sex ratios or adult sex ratios of many vertebrates show greater diviations, however. In the Alaskan fur seal one adult bull may dominate a breeding group of 30 females. Wheras the total adult male sex ration is greater than 3 percent the excess adult males are on the periphery of the breeding grounds, they are essentially surplus, and they suffer substa-ntially higher mortality. In most primate populations, adult females out number adult males two or three to one.

Apparently males have higher mortality rates throughout life than females have higher mortality rates than males, particularly during the nesting season. In waterfowl and gallinaceous birds, for example male often outnumber females. The sex ratio of blue-winged seal (Querquedula discors) in fall and winter flocks averaged 59 percent males in a sample of 5,090 birds and in the early summer breeding population of mallards (Anas platyrhynchos) and pintail (Anas acuta) in Manitoba, males may outnumber females 4 or 5 to 1.

Life Tables

Life tables represent tabular data on age structures of populations, and they also provide information on mortality rates and longvity patterns. The essential data of a life table may be of two basic types: (1) census data on a population with accurate counts of the numbers of individuals in each age category, (2)mortality data on the number of individuals of each age group dying in a given period of years. From such data, surviviorship and longevity can be calculated Insurance companies use life table statistics on human populations to estimate the probability or death at each age group, and from these data they establish rates for life insurance. Life tables provide ecologists with essential information for the management of animal population. They pinpoint mortality patterns, indicating when deaths occur and how many. If one finds abnormally high mortality occurring in infancy, adolescence, or early adulthood, it may suggest specific management procedures to alleviate this mortality.

Excessive infant mortality, for example, may be associated with inadequate food for young animals, or inadequate cover by which they can escape predators, or improper patterns of parental care due to behavioral disturbances within the adult population. Excessive adult mortality might indicate shortage of adult food supplies, infectious diseases or parasitic burdens which become more severe in adulthood, or possibly excessive crowding during the reproductive period so that sexual fighting becomes violent. Thus, life data provide important clues for more detailed study. They seldom give the final answers on why or how certain events occur, but they form the essential starting point of what is happening in a population.

Populations may be studied in both anatomical and physiological terms. That is, we may analyze the structure of a population in terms of numbers, density, spatial distribution, sex and age ratios, and other descriptive qualities. Similarly, we may analyze patterns, movements, and other dynamic attributes. The ultimate objectives are to understand the organization of populations and their vital processes.

The question of population has occupied the attention of state men and philosophers since ancient times. But it is only recently that attempts have been made to investigate systematically the cause of population change and the specific ways in which population dynamics influence human welfare. Plato and Aristotle were interested in the question of population size in the context of a city-state. To them the optimum size of the population was that is with which man's potentialities could be fully developed and his "highest good" realized. Thus was possible when the population range was enough to be economically self-sufficient and capable of defending itself, but not too large for constitutional government. Plato specified 5,040 as the number of citizens "most likely to be useful to all cities".

European authors on population during the early modern and the medieval periods favoured a growing population. The discovery of the New World, the increase of commerce between Europe and Asia, and the rise of national state brought about some changes in the terms of discussion of population questions, but not until the later part of the eighteenth century was there any marked shift in the general attitude favouring large and increasing population.

The mercantilist and cameralist schools of political economy, which flourished in Europe during much of the seventeenth and the eighteenth century, emphasized the economic, political and military advantages of large and growing population and favoured various measures to stimulate population growth. These writers were concerned primarily with the ways and means of increasing the wealth and power of the State. Their aim was not to promote per capita income but to increase the aggregate national income which was viewed as a sources of revenue for the State. During the last half of the eighteenth century, a number of writers on economic and social questions rejected the mercantilist doctrine and the view that population growth was advantageous and should be actively encouraged by the State.

Some writers, specially in England, France and Italy, emphasised the dependence of population upon means of subsistence. They opposed arrangement for poor relief on the ground that it might undermine frugality, make labour immobile, depress productivity and thus increase the pressure of number on means of subsistence. These arguments were turned against such advocates of social reform as Godwin and Condorect, in an effort to show that any benefits from reform would be cancelled by a consequent increase of populations. At the close of eighteenth century, Thomas Malthus, an English economist, noted this tendency for populations, including human population to increase rapidly.

In 1798, he published his famous work, Essay on the principal of population, in which he put forth the idea that populations tend to increase faster than their means of subsistence. In general, he felt that populations tended to increase geometrically or exponentially, whereas their food supplies and means of subsistence tended to increase only arithmetically.

A geometric increase is a series of numbers having a common ratio, such as 2...4...8...16...32...64 in which each succeeding number is twice the former. An arithmetic increase is a series of number having a common difference, such as 2...4...6 ..8...10...12, in which each number differs from the other by 2. Thus, Malthus thought that populations constantly tended to outstrip their food supplies and were then decimated by poverty, starvation, disease, warfare, or other catastrophes. Malthus was primarily concerned with human populations, and he felt that moral restraint was the solution to human population pressures. Malthus's theory is thus based on two postulates and one assumption. His postulates are that (i) food is necessary for the existence of man, and (ii) that passion between sexes is necessary, and will remain nearly in its present form. The assumption is that the production of food increases in arithmetical progression while the population increases in geometrical progression. Thus the power of population is indefinitely greater than the power of the earth to produce subsistence for men. Efforts must, therefore, be made to keep the population from growing beyond the means of subsistence through some powerful checks, else it will lead to "vice and misery."

Malthus talked of two kinds of checks to population positive and preventive. He termed preventive check as "prudence" and included in it postponement of marriage and restraint on reproduction. Positive checks he termed as "natural" as "they arise form the situation itself" and included in it wars, pestilence, epidemics, diseases, famines, natural calamities and the like. Malthus was highly praised and also bitterly criticised for his views. His Essay aroused a storm of controversy which long out-lived Malthus himself, and which made both his followers and his opponents conscious of the need for adequate information on population trends and their influence on social and economic conditions. Thus Malthus was indirectly responsible for developing population censuses a vital statistics.

Malthus has been criticised for his postulates, his assumptions and his conclusions. While it is not to be disputed that food is necessary for man, but so is water clothes, shelter, and such other prime necessities of life. There is enough historical evidence to indicate that the output of food and of other necessities has increased faster than population, and has been brought about without any distinct symptoms of increasing "misery and vice", but rather of their diminution. With improvement in techniques of production increasing returns are being obtained from land and from the production of prime necessities of life.

Malthus has proved a false prophet. While he postulated great miseries on account of unprecedented growth of population, it is found that birth rate in most of the advanced countries of the world has declined, and that the rapid growth of population in under-developed countries is basically due to a rapid fall in death rate and not on account, as Malthus feared, of an increase in birth rate. Marx rejected the Malthusion theory of population. He held that there is no universal law of population, and that the source of "over-population" was not to be found in man's biological power of reproduction, but in prevailing capitalistic mode of production. Over population arose because capital accumulated less rapidly than did the supply of labouring population was not only a necessary consequence of capitalistic accumulation, but also a necessary condition for the existence and continuation of the capitalistic form of economic organisation. The realisation that Malthus has over-generalised a particular case lead to a re-thinking of the population question. Writers at the beginning of the present century began thinking that population growth is not always undesirable. This led to the development of the optimum theory of population which is associated with the name of Professor Cannan and others.

The theory says that given the natural resources and the technique of production, that population, Cannan envisaged a point of maximum return for all industries taken together, that is the population which will maximise productivity. He also indicated that the optimum size of the population changes as circumstances change and as new techniques of production are introduced. Effort to precisely define the "optimum" led to the refinement of the concept, but at the same time to the theory being criticised as of little practical value. It is considered highly artificial to select "average real income "as the soletest of the optimum. In fact there can be an optimum size of the population from ecological, political and other points of view. Again, the optimum size is subject to frequent changes on account of changes in techniques of production.

The transition theory, development in the 30's of third century and associated with names of Warmn S. Thompson and Frank W. Notestein, explains the relationship between population with and economic development. It says that in developing countries, which are characterised by heavy pressure of population, low productivity heavy dependence on agriculture, rudimentary techniques of production, under-developing means of transport, and poor sanitary conditions, birth and death rate high.

Death rates are high because of poor diet, primitive sanitation, and the absence of preventive and curative medical facilities. Birth rates are high because of social beliefs and customs which encourage large families and which group up if a high death-rate community is to continue its existence. The birth rate around 40 per one thousand person and the death rate around 35 so that the population does not grow at a fast rate. With economic progress the death rate begins to decline. This is because of improved means of transport, better sewage disposal and improvement in drinking water facilities. But the birth rate stays high so that the difference population tend to grow at the rate of 20-30 per thousand population per year. Because of the rapid rate at which the population grows, this period is also known as that of "population explosion". One of the features of economic development is typically increasing urbanisation and in an urban setting children are usually more of the burden and less of an asset. The processed economic change also weakens the force of traditional customs and beliefs. Females are being to realize that if burdened with a large family they are unlikely to play their due role in society.

Consequently, the idea of large family yields place to the small-family-death and the birth rate declines from the high 40 to about 17 per one thousand population. The death rate is also low, about 8 per one thousand, so that the population just maintains itself. The three stages described above are also called the stages of slow population growth, rapid population growth and of stable or declining population growth. The industrially advanced countries of the world have passed through the first-two stages and are current in third stage. Developing countries of Asia, Africa and Latin America are either in the first stage or entering the second stage. The sequence of events described by the theory can be traced in every region when the economy has evolved from an agrarian to an industrial market oriented economy. But it is not specific to quantity if the degree of decline of the vital rates. The theory, however, contains one significant generalisation, namely, the decline in birth rate occur after a substantial time-lag in comparing with the decline in death rate, and that during this lag population increases very rapidly.

For instance, the "area of European settlement" increased its population six fold during the period 1750 and 1950. The population more than doubled in the century from 1750 to 1850 and nearly tripled in the period 1850 to 1950. The theory is of considerable significance to under developed countires of Asia, Africa and Latin America. Their birth rate is high and the field of public health, it has been possible for the death rate to be substantially reduced with marked changes in the economy or in the birth rate.

Consequently, their population is growing at a rate which doubles itself in twenty to twenty-five years. This is tending to retard economic progress. There is therefore an urgent need for bringing about a substantial decline in birth rates if economic development is to become a continuing process.

The first major scientific challenge to Malthusian doctrine as a pattern of population growth appeared in the 1830's in the work of Pierre Verhulst. In 1839 Verhulst proposed that population normally grow in a much more orderly fashion than that proposed by Malthus, describing in their growth a curve of S-shaped proportion. This became known as the Logistic Theory of population growth.

The logistic theory asserted that population have a slow initial growth rate, which increases exponentially unitl it reaches a maximum, and then becomes progressively less as the population approaches an upper limit of its growth. The upper limit is approached gradually and in an orderly, predictable manner. The resulting curve is the familiar S-shaped of population growth. It should be noted that the logistic curve and the Malthusian or irrupting curve do not differ in the early stages of population growth—both show a slow start followed by a period of exponential or geometric growth but they· do differ fundamentally in the upper or controlling stages of growth. The Malthusian curve is characterized by an erratic, often catastrophic, pattern of limiting growth, whereas the logistic curve characterized by a smooth, orderly and gradual pattern.

Following the work of Verhulst, the logistic theory lay dormant unto it was independently. derived and popularized by Raymond Pearl in the 1920's. Pearl applied the logistic curve to the population growth of yeast, Drosophila and man. This ushered forth a burst of activity resulting in the application of the logistic curve to the population growth of protozoa, water fleas, pond snails, thrips, ants, bees and other organisms by many investigators.

With this wealth supporting evidence, the logistic thoery attained wide acceptance. It was proclaimed as a law of population growth by some biologists and was used to predict future population levels of experimental and natural population. Pearl's investigation led him to conclude: "It has been shown in what has preceded that populations of living thing `from the simplest' represented by yeast, to the highest, represented by man, grow in accordance with the logistic curve. One can now feel more certain that this curve is a first and tolerable close approximation to a real law of growth for human populations.... We can, for example, upon a more adequate scientific basis mere than guess-work, predict future population, or estimate past populations, outside the range of known census counts."

That some biologist accepted Pearl's interpretation rather freely is indicated by statements of Clarke (1954): "Population of a wide variety of organisms, ranging from bacteria to whales, have been found to follow the logistic curve in the growth from.... The growth of man's population follows a similar pattern whether examined in individual regions or in the world as a whole. "Clarke presented a graph of population growth in the United States fitted with a logistic curve, and observed an upper symptote of 184 million in year 2100 A.D.

Many other biologists and mathematicians, however, severely criticized the significance of the logistic theory. In fact, the theory met stern opposition since its first expression could be made to fit a wide variety of curves, often of very different nature. Gray, Hongben, and Wilson and Puffer registered doubt as to the universal validity of ligistic theory and commented on its inadequacies as a description.

Feller pointed out that the actual fit of the logistic curve to much of the published data was poor and better agreement could be obtained with other formulas. Andrewartha and Birch and sang accurately pointed to limitations of the logistic equation. Sang in careful reexamination of population growth in Drosophila, emphasized that, "The data we have summarized show that only in very exceptional circumstances would logistic growth occur in a Drosophila culture... The ecological situation is too complex to be adequately described as the Pear-Verhulst law, "Sang continued to discuss general aspect of the logistic theory by stating, "This acceptance of a plausible formula, in spite of its inaccuracy, may be justified; provided this attitude does not inhibit future research, particularly in to fundamental conception. In the case reviewed, it seems that wide acceptance of the logistic law has led to just this kind of inhibition of further work, and that many interesting ecological and physiological processes have been ignored as a result. The study of vertebrate population in both natural and confined circumstances has produced some population growth curves deviating considerably for logistic growth and tending to ward errupting or Malthusian growth. The sheep population of southern Australia showeed a pattern of liogistic-like growth from 1840 to 1890, but its upper level was reached abruptly and was followed by substantial mortality. The deer of Kaibab National Forest in Arizona grew from an estimated population of 4,000 animals in 1918 after a period of extensive predator control to an estimated population of 1,00,000 in 1924 but in the winter of 1924, they died by the thousands after the population had destroyed its food supply. Their growth and decline described a classic Malthusian curve.

The pheasant population of Protection Island, Washington, showed a phenomenal increase from 8 birds on the 397 acre island in 1937 to nearly 2,000 birds by 1942. In one year the population showed an increase of 3000 percent. Unfortunately the field research on this population was terminated before natural control occurred, but the population seemed to be describing a typical irrupting population curve.

Several studies of seminatural populations of wild house mice have also shown irrupting population curves. Seminatural populations have been maintained in confined, but free-ranging conditions with a natural habit for this species, except for the element of confinement. Various factors such as space, food supply, resting cover and the structure of the physical environment have been experimentally controlled to elucidate their roles in population growth and limitation Work of this type at the University of Wiconsin and the Institute of Ecology in Warsaw Poland, has shown that population growth in *house mice* is highly variable, and often dependent upon patterns of sicial behaviour within the population. For example, populations in identical environments showed widely variable patterns of population growth. The variability was related to social behaviour within the population in terms of aggressive behavior, dominance hierarchies, sexual behaviour and parental care.

All populations become controlled even though abundant food and nesting space was still available when aggressive behavior increased to an abnormal degree, when the nesting territories of females were no longer properly maintained, and when parental care of the nestlings become aberrant. Infant mortality increased through nest desertion and cannibalism.

The densities at which these behavioral pathologies occurred were highly variable between populations, and they were function of the crowding tolerance and behavioral stability of individuals. This complicated interplay of ecology and social behavior was not adequately accounted for by the logistic theory and the resulting population curves were totally individualistic and unpredictable. A fairly complete evaluation of logistic theory has been presented by Andrewartha and Birch. They have analyzed the inherent assumptions of the logistic equation and have reviewed many studies in the light of these assumptions.

The analysis has led them to conclude that the logistic curve provides an inperfect, yet useful, description of the population growth of organisms with simple life histories (such as yeast and Paramecium); but for animals with complex life histories (such as insects and vertebrates) the population growth usually does not conform closely to the logistic curve except for small parts of the data. The concluding attitude of Andrewartha and Birch is well experessed in the following. "despite its theoretical limitations, the logistc curve remains a useful tool for the ecologist. Because of its limitations, too much reliance should not be placed on it in particular cases until it has been verified empirically for each case. "Hence, it appears fallacious to utilize the logistic equation of a prognostic purpose in any population without experimental precedence. The evidence is clear, therefore, to support the early contention of Wilson and Puffer that the logistic formula should not be considered a law of population growth to such an extent that it permits extrapolation of the curve of forecasting purposes or the interpretation that the constants of the equation are the constants for nature. Rather, it does provide a convenient description of growth in certain well-studied populations.

The major question in human terms is, of course, what pattern of population growth characterizes man? From the early seventeenth century to the present time, the world's population has displayed exponential growth. There seems to be widespread faith that human populations shall ultimately reveal a logistic pattern, but certainly all attempts to predict human population levels by means of the logistic model have been very far from the mark. Pearl predicted a world population of 2.6 million by the year 2100.

This population level was exceeded many years ago, and economic planning based on Pearl's prediction would indeed have been disastrous. Although we continue to hope that human population throughout the world will achieve a pattern of graduallogistic balance through orderly and intentional process, there is no sound scientific evidence that this will in fact occur. The possible of catastrophic mortality through war, disease and starvation are still very much with us. So also are other patterns of mortality which are not necessarily catastrophic in their total effect, but are clearly related to density, crowding and modern population pressures.

In this category are automobile accident mortalities, which are responsible for 50,000 deaths per year in the United States alone, various stress-related diseases, including coronary artery disease and high blood pressure, and various pollution-related diseases including emphysema and chronic bronchitis. If deaths attributable to these causes continue to show present trends, these mortality factors will begin to play a more significant role in population regulation, especially in technologically advanced countries.

Different patterns of population growth may be exhibited by populations recovering from major depletions, or by populations invading new inhabitants. Since most populations are already established, however, and are not growing de novo, they are instead showing varying degrees of fluctuations.

Our familiar concept of the balance of nature implies stability but the fact is that most animal populations under go continual change. Natural populations often exhibit a dynamic resiliency, rising and falling in response to many factors. In one sense, a pattern of population fluctuation such as that normally exhibited by most species, may be considered continuous series of growth & decline sequeuces. Thus study of population fluctuation becomes closely allied with the study of population growth and natural control.

At the outset it is convenient to separate population fluctuations which are seasonal (i.e., obviously related to seasonal weather), from those which are nonseasonal latitudes, most animal pupulations have breeding seasons so that population growth typically occurs in the spring and summer seasons. Most arthropods and vertebrates begin producing young in the spring and summer seasons, so that these seasons are characterized by population growth, and they cease the production of young in laten autumn and winter. In aquatic ecosystems, many populations also show marked seasonal fluctu..tions. Both phytoplankton (miscoroscopic algae), and zooplankton (protozoa rotifiers, small crustacea, etc.) usually show spring and fall increases in population. These sharp increases, often 10 to 20 fold, are known as "pulses". They are sometimes correlated with temperature changes or with turnovers in the water state which recirculate nutrients.

In the tropical regions of the world, where there are not sharply defined seasons on a temperature basis, reproductive seasonality still occurs in many plants and animals, and it is often related to the cycle of wet and dry seasons. Many tropical insects, for example, have sharp peaks of abundance related to the monsoon seasons. Even vertebrate animals, including primates, have been found in nature to have definite periods of breeding and birth seasons, often correlated with the monsoons.

In both tropic and temperate zones an important principle seems to be that the young appear at the most favourable time of the year in term of food and climate. Thus young deer and antelope are born when insect food is most readily available. These are exceptions to this general principle; for example, young rhesus monkeys are born in the midst of the hot dry season in northern India, when temperatures are excessively high (often over 100 degree F), water is extremely scarce, and plant growth is virtually nil. In this case, however, the infant feeds exclusively on the mother's milk for two or three months until the monsoon season starts.

It is possible that natural selection has acted in such a way as to place breeding, conception and the first trimester of pregnancy at the most favorable time of the year, namely, the immediate postmonsoon month when food is most abundant. In other words, the events of conception and early embryonic growth may have been more critical in the life cycle of the rhesus monkey in terms of evironmental influences than the period of early infancy which is well protected by the mother and the social group. This is speculation and not established fact at this time. It also fits with the observation, however, that the monsoon season with its luxuriant growth again arrives when the infant is two or three months old and is relying less on maternal milk and starts taking its own food. Population fluctuations which are relatively independent of seasons are of two general types: (1) random, or (2) cyclic. Cyclic fluctuations are those which have a definite periodicity, or rhythmicity; that is, the population reaches a peak at fairly regular intervals. In vertebrate populations, certain species reach population peaks every 4 years, while other have a 10-year cycle. These will be discussed later as specialized cases of more generalized random fluctuations.

Ramdom fluctuations may be minor perturbations of fairly stable population, producting relatively flat curves, or they may be major changes in abundance, producing erratic unstable curves. Examples of stable populations are increasingly difficult to find in the modern world. They were probably characteristic of many organisms in undisturbed ecosystems; that is, ecosystems in balance. They were also probably characteristic of complex systems, such as forests and particularly tropical forests where great species diversity produced a complicated web of natural check and balances on each species. Within such a system most vertebrate populations were probably stable and exhibited only their typical seasonal fluctuations Probably some of the best modern examples of stable populations may be found among the birds and larger mammals. In birds where considerable species diversity occurs and fairly extensive food resources are available, considerable stability in population size may be present.

John Gibb reviewed a number of bird population studies several years ago and noted: "Notwithstanding the considerable annual fluctuation is bird populations that have been recorded in almost every species investigated, it is their comparative stability over long periods that is their most remarkable attribute. "Thus, although wide variations is number occur form year to year in many bird population, these population often show a tendency to return to typical levels. An example of a relatively stable vertebrate population is afforded by a census study of rhesus monkey populations (Macaca mulatta) in northern India. With complete protection, good health and abundant food supplies, rhesus populations can increase in experimental colonies at the rate of 15 per cent per year. But in India, where monkey experience limited food supplies, varying degrees of protection from the human population, and a considerable number of infections diseases, populations do not show this pattern of increase.

Throughout India as in most of the world, primate population are declining due to habitat deterioration (cutting of forest, overgrazing of savannah lands, etc.) human competition (where native people shoot monkeys for food), and high rates of trapping for commercial trade. As an example of the latter, in the late 1950's India exported more than 100,000 monkeys per year for use in biomedical research and pharmaceutical production, though the current figure is only about onehalf of this. These monkeys were essential in the development of polio vaccine, measeles vaccine, and many types of medical and psychological research. The long term population trend of many animals, especially wild vertebrate animals, is down ward. Since 1600 A.D., 120 species of birds and mammals have become extinct, a number several times greater than the natural rate of extinction.

At the present time, approximately one vertebrate animal become extinct on the face of the earth every year, and within the next 30 to 50 years, 100 more species will probably be lost. The International Union of the Conservation of Nature periodically publishes a list of rare and endangered species (the publication is known as the Red Data Book), and it now lists nearly 300 species of birds and 275 species of mammals as endangered. The peregrine falcon (Falco peregrinus), ospery (Pandion haliaetus), and bald eagle (Haliaectus leuocephalus) are presently among the vanishing animals of North America. Others have come close to extinction, but were saved by intensive conservation movement at the last moment. Throughout the world, the basic causes of extinction are habitat deterioration, poaching, poisoning, and direct or indirect competition with man. From the ecological standpoint, some extinction is an inevitable evolutionary process, but when it becomes excessive as in the last century, it is a serious disruption to biotic communities and ecosystems. Each extinction lessens species diversity, and thereby reduces ecosystem stability. Each estinction also represents the irreplaceable loss of unique biological material.

Some vertebrate animals have increased drastically within the last century, but these are usually domestic animals such as dogs, cats, cattle, goats and so forth, or wild animals that have learned to become commensal with man; that is, animals which have successfully adapted and even exploited the human environment. This includes the domestic rats and mice such as the Norway rat, the house mouse the Asian bandicoot rat (Bandicota bengalensis),the Asian house shrew (Suncus murina) to starling, the house sparrow and the pigeon

(Columba livea.) Even the herring gull (Larus argentatus) is adapting to man's urban environment successfully by feeding on garbage dumps, and gull population on both United States coasts have increased in recent years. Apart from those species which seem to be headed towards a permanent decline and eventual extinction, and those which thrive on man-made environments, the fact remains that most animal population fluctuate substantially from year to year. It is often a matter of common knowledge among farmers and sportsmen that "this is a good year for pheasants,"or "there were rabbits all over the place last year, but not so many this year," or "this is the world year for mosquitoes we've ever had," and so forth. These are all popular expressions of population fluctuations.

Scientifically, these fluctuations are not often recorded in a systematic and continuous fashion, but some data do exist to document the types of fluctuations which are common in animal populations. In trying to manage animal populations, we tend to do the most obvious thing, which would logically seem to produce results, and often does, but in the long run, may be quite detrimental to the total ecological balance of a community.

Thus, in controlling insect populations, we use residual poisons such as DDT, but are now discovering after some 30 years, that this has long term side effects which are quite dangerous, and we are also dicovering that many insects have become resistant to it in any case. In controlling the population levels and general health of their prey species. In enhancing the abundance of desirable species, we have often poured money into artificial breeding and restocking programes of animals such as pheasant, wild turkeys and trout only to find that in the meantime, the habitat has so deteriorated through pollution that the transplants have little or no change of success, and the recovery rate per transplant is almost nil. This is not intended as a blanket condemnation of all of our control and management programs, but by and large the balance sheet is not favourable, mainly because we have based our management on trial and error rather than on sound basic knowledge of the ecological factors involved in population growth and regulation.

The most diamatic population fluctuations are those which are cyclic: that is, in which the population reaches peak numbers at fairly regular intervals. Some animal populations, such as those of the snowshoe hare (Lepus americana) have periodicity of abundance of 9 to 10 years. Other animal populations have a cyclical periodicity of 3 to 4 years. These include the Norway lemming of Scandinavia (Lemmus lemmus), the field mouse (Microtus agrestis) of Europe, and some of the predators of these two notably the colored fox (Vulpes fulva) and the snowy owl (Nyctea scandiaca). Not all populations of these animals always fluctuate cyclically. When they appear to do so, sophisticated statistics may be required for proof of real cyclic periodicity.

These animal population cycles have been intriguing and puzzling to biologists for many years. Considerable research has been centered around trying to understand the cause of cycles, and numerous theories have been proposed to account for them. The final answers are still not clear, though many biological phenomena have been discovered and analyzed in this search for understanding.

Two types of factors have been considered as the basic causes of animal population cycles:(1) extrinsic factors; that is, those outside of the population itself, such as climate and weather, cosmic events competition between species, predator-prey relationship's and food supply; or (2) intrinsic factors; that is those within the animals as population increases occur and they become crowded.

Several of these factors can be dismissed rather quickly, since extenive research has failed to show any clear relationship. For example climatic and meteorological changes have not been satisfactorily correlated with regular long-term cycles of population change in any animal. Sometimes population peaks coincide with years of great rainfall, or some other notable aspect of weather, but these occurrences are by no means consistent. At one time, sunspot cycles were considered to be correlated with population cycles. Sunspot maixma have occurred every 10 to 11 years, and they were correlated with peak population of hares and lynx in the late eighteenth and early ninteenthe century, but in the late eighteeth and early ninteenth century, but in the last 100 years, have been entirely out of phase.

Changes of populations of predators such as the lynx, fox and sowy owl are clearly correlated with cycles of their prey. But they cannot be shown to cause the prey cycles, for prey cycles often occur in areas devoid of predators. For example snowshoe hare populations have cycles in areas where no lynx are found. There has been some attention given to disease and parastism as the cause of long-term cycles.

8 Population Interaction

There are two behavioural phenomena which are widespread throughout the animal kingdom, and which have considerable significance in population ecology, are territorialism and dominance hierarchies.

Territorialism is essentially a social pattern of spatial utilization by which individual or groups have control over certain units of space.

Dominance hierarchies are an-ordered systems which determine individual access and priorities to natural resources. Both of these phenomena relate to population ecology intimately because they represent systems of behavioural control which affect the abundance and distribution of animals, their reproductive patterns and their mortality patterns.

Territorialism

A territory has been defined as any area which is defended against other member of the same species. It may be occupied and defended by an individual as in the case of a male stickleback fish (Gasterosteus aculeatus); by a pair as in the case of many birds or by a social group, as in the case of vervet monkeys (Cercopithecus aethiops) and gibbons (Hylobates lar). It may encompass most or all of the occupant's home range including his nesting and foraging areas, as in primate territories; it may include only a limited area around a nest site of feeding site, as in many birds; or it may involve only a very small plot of ground on which mating occurs, as in the Uganda kob (Adenota kob), an African antelope. Hence, territorialism is a highly variable and complex phenomenon, and involves a wide spectrum of behaviour patterns.

In all cases, however the essential feature is that individual animals, or groups of animals, have controlling ownership over a given piece of property. Within this property they have the right of utilization over the space and the resources contained therein. Territories are established and maintained by a great variety of behaviour patterns and communicative displays. Initially, the establishment of a territory may involve active aggressive behaviour and fighting.

The territorial combats of siamese fighting fish, jungle fowl, elephant seals and wild antelope are well known. Those combats are often vigorous and are true tests of strength, injuries are common, and in rare cases, death may occur. Most frequently, however, the vanquished animal merely retreats and vacates the territory. Once established territories are most often maintained by communicative signals and displays. These signals may be visual, auditory or olfactory in nature. Most fish have visual territorial displays which consist of spreading the fins and brightening the colors. Some fish have active sound production in territorial calls, as for example the Atlantic toadfish (Opsanus tau). Most birds which are territorial have a combination of vocalizations and visual displays. The male song bird of many species perches in a conspicuous spot and repeatedly gives forth his song, which serves more often as territorial threat than as a mating attraction. Mammals typically use a combination of auditory and olfactory signals as communicative displays in the maintenance of territory. Some of the arboreal primates, such as the gibbon and the howler monkey (Alouatta palliata) have an advanced vocal pattern which conveys information about the position of the group and indicates that the territory is occupied.

Many carnivores, such as the wolf, lion and tiger (Felis tigris) have territorial calls which can be heard for several miles, and they also have olfactory signals in scent glands and urine which are used to mark the boundaries of territories. The scent signals may persist for days or even weeks, and thus serve as more stable cues than serve as more stable cues than vocalizations which fade rapidly.

Territorialism is common among many higher invertebrates (crustacea, insects, spiders, etc.) and throughout all vertebrate classes. Territorialism can be found in fish amphibia, reptiles, birds and mammals, which seem to operate on much the same principles and have basically the same functions. Although this is true, it must not obscure the fact that there are many animals which do not show territorialism.

Many species do not stake out specific territorial claims and do not actively defend given tracts of habitat. Often one can find closely related species which are territorial in one case, and nonterritorial in another. In ungulates, for example, the Indian blackbuck (Antilope cervicapra) is territorial, whereas the Indian swamp deer (Cervus duvacelli) is not. In primates, the gibbon and the howler monkey are clearly territorial, whereas the rhesus monkey (Macaca mulatta) and the chimpanzee (Pan troglodytes) are not. Even within the same species there are circumstances in which the individuals of some populations are territorial distributed and other circumstance in which they are not. This is true in ungulates for example, many of which are territorial during the rut or mating season, but the same individuals are non-territorial at other times of the year.

Apart from this obvious correlation with breeding, however, there is evidence in some species that territorialism is related to the population density and/or environmental circumstances. In the Japanese salmon (Plecoglossus altiveli), individual fish are territorial at low population density, but as density increases, individuals give up their territories and form a school.

Wild house mice (Mus musculus) show similar behaviour with territorialism at low population densities, and mass group behaviour at high population densities. This variability is related to environmental factors and differences in social behaviour of key individuals. In view of these facts, it does not seem warranted to consider territorialism an instinctive pattern of behaviour. Territorialism is certainly common and wide spread, but by no means universal within limited animal groups or even within a single species.

Dominance Hierarchies

Dominant animals have priority access to food, mates, nest sites and resting locations. Dominance is often manifested between animals by single physical displacement by the displacement of one individual by another at a feeding or resting site. Occasionally it involves displays or threats, and rarely it may involve direct fighting. The latter is more common in initial encounters in which the dominance rank or social status of each individual is not yet determined, but once established the hierarchy is then usually established by display or by social memory.

Once a rhesus monkey has established high rank, it does not have to constantly display this rank in a stable social group. The high rank of some individuals and low rank of other become an accepted social norm within the group. Only when the situation becomes unstable through death, emigration or invasion does there have to be an avert re-establishment of rank within a dominance hierarchy. So far as we have seen territorialism and dominance hierarchies as two types social systems which play a regulatory role in population dynamics, the first by regulating of space second by controlling individual priorities within a common space.

In some animal species and populations, territorialism and dominance hierarchies apparently serve as sensitive feedback mechanisms which influence dispersal, mortality reproductive success and population size. In other species and circumstances, territorialism and hierarchical structures are often subject to breakdown and collapse.

Ecological consequences are then profound and extensive. In many rodent populations, excessive population growth and crowding is often associated with the following phenomena: a breakdown of stable territoria-lism and dominance structure, reduced fertility as evidence by lowered pregnancy rates, high prenatal loss and low infant survival, poor nest construction and poor maintenance, nest fouling and pollution, high aggression and wounding deteriorating health in the majority of the population, abnormal aggregations of individuals in the 'behavioural sink' pattern and finally population decline.

The detailed study to territorilism and dominance hierachies is really in the realm of ethology and animal behaviour, but the functions and consequences of these behavioural systems have a great deal of ecological significance. It has been abundantly demonstrated in many territorial animals that the holders of territory have the greatest breeding success. In fact, in some animals, non-territorial animals usually fail to breed. Similarly in those social groups characterized by dominance hierarchies the high ranking animals have the greatest breeding success and the low ranking members have less success or fail entirely in breeding efforts. In chicken flocks, low ranking roosters show frequent counting behaviour, sometimes more than dominant rooster, but they have relatively little success in completing copulations. It thus seems clear that the social systems of territorialism and dominance hierarchies have a limiting function in population dynamics; that is, they limit the breeding to those individuals which are ecologically and behaviorally most successful in establishing territory or high rank.

Theoretically, as population densities increase, and space becomes more limited, a smaller proportion of individuals within the population is able to achieve territories or high rank, and thus total reproduction is reduced in relation to population size. Ideally, these behaviors are density-dependent population limiting factors; that is, the higher the density, the greater the limiting effect on the population. The role of social behavior in population regulations has been discussed at great length by Wynne-Edwards.

Wynne-Edwards proposed that the basic long-term population level of many animals is established by food and resource availability, but that social behaviour establishes a more immediate regulating influence on population size and density. This influence on social behaviour, he felt, comes into play before critical population levels are achieved; that is, before starvation and/or damaging violent aggression are conspicuous. Thus, according to Wynne-Edwards most animals have intrinsic, self-regulatory population mechanisms which operate via social and behavioural feedback systems.

These social and behavioral feedbacks are usually ritualized displays which relate to territorialism and dominance hierarchies, and thus convey information about population density.

Therefore, regulatory feedback is provided according to the intensity of social competition. In other words, ritualized displays, or conventionalized rivalries, regulated breeding success and thus population numbers. Wynne-Edwards provides numerous examples of this hypothesis, including the complex display patterns of gallinaceous birds, (prairie chickens, sharp-tailed grouse, peacocks, bower birds et al.) waterfowl and many song birds. He also considers the vocal choruses of amphibians, the schooling of fish, and the flocking and birds and mammals as examples of complex social behaviour which serve the function of conveying information about population density and social competition.

Many aspects of Wynne-Edwards' theory are controversial especially when he theorized that most social behaviour has actually evolved as a means of controlling population levels. Wynne-Edwards proposed, for example, that territorialism has arisen as a social means of preventing over population. This theory has met substantial criticism from other ecologists and students of behaviour.

These authors felt that there is little reliable evidence to support this hypothesis even though it is logically very attractive. Brown pointed out that, "Population regulation is never completely under the control of the species by itself, but depends in a complex way on interactions between members of the ecological community." Thus, Brown felt one must consider the interaction of all factors, including predation, disease, food supply, interspecific dynamics of any given species.

Brown found satisfactory evidence that territorialism tends to space individuals over available habitat, that it increases emigration and morality rates in individuals unsuccessful in finding territories, and that it has arisen in response to individual selection resulting from aggressive competition. The more detailed aspects of this controversy are out side the scope of introductory ecology.

On the more basic parts of the hypotheses—namely, that social behaviour provides some regulatory feedback on population ecology—there is certainly no question that this is true in most animal populations. Even this concession must not obscure the fact, however, that some animals have such poorly developed and/or variable responses to increased density and crowding that they can readily breed themselves into starvation or conditions of violent aggression and extreme social strife.

White-tailed deer populations, for example have frequently increased to the point of extensive starvation. Lemming and other rodent populations have also increased to the point of mass behavioral strife where social mortality becomes extensive. Quite probably, some human populations are also in this category, providing example of populations in which social feedback limiting population growth often does not into operation until the point of starvation arrives (as in China and India), or until the point of violent and destruction aggression breaks forth (as in many wards in which population pressure has had a causative influence) or until social morality becomes limiting (as in countries where abortion is a significant form of population regulation).

It is scientifically important to consider animal population in which environmental crises and/or behavioural breakdowns occur. In terms of human value judgments these seem to be cases of population regulation by wasteful and sometimes tragic means. In evolutionary terms, however, many of the populations showing these phenomena are of highly exploitative natures, meet new environmentral situations had expand into available opportunities. Perhaps their lack of frigid, stereotyped behaviour and ecology, and their lack of precisely tuned population feedback systems, enables them to seize each new ecologic opportunity to the fullest extent. Some of the best examples of such species are the domestic rodents especially the wild house mouse, the wild Norway rat (Pattus norvegious) and the Asian bandicoot rat (Bandicota bengalensis). Numerous ecological studies over the last 25 years have shown these species to have the following population characteristics:

1. They have tremendous reproductive potential and under certain environmental circumstances they often show invasive and eruptive population behaviour.
2. Populations can readily exceed the carrying capacity of the environment and this can result in sudden population decline.
3. Populations fluctuate widely, natural control is often abrupt and "ecologically wasteful", in terms of individual health and welfare.
4. Population regulation often involves social disruption and destructive behaviour.

These characteristics have been observed frequently in both natural and artificial populations of rodents over the past 40 years. The natural social behaviour of the Norway rat and wild house mouse includes both territorial and hierarchical behaviour. Females tend to the territorial around the nest site so that they normally exclude adult males and strange females. Often two of three females nest together in a communal agreement. Males are also territorial if population density is not excessively high, and high ranking males defend a system of runway and the females living therein.

The size and patterning of the territories depend upon the rank of the male, the density and social structure of the population and the nature of the habitat. In a stable habitat both population levels and social structures of domestic rats might remain quite stable over a period of time. But the habitats of domestic rodents are characteristically unstable with changing amounts of food (around farm, for example where harvests and food stores come and go), changing mortality patterns (as control compaigns) or disease outbreaks wax and wane) or changing cover conditions(as trash availability rises and falls) so that the populations are often experiencing drastically changing resource conditions. Under these circumstances, the populations respond rapidly with surges or declines of growth.

Under periods of rapid population growth leading to high densities, a variety of phenomena have been observed. There is often rapidly increasing social contact resulting in an increase in aggressive interactions. This frequently lead to a breakdown of territorial boundaries, increased crowding of the nest sites and sharply increased mortality of young nests become trampled and destroyed. Females often desert the nests or actually kill the young. Wounds become frequent, especially an adult males forming sites for mites to invade the skin, so that acarine dermatits may achieve epidemic proportions. With a breakdown of territorial behaviour, bizarre aggregations developed so that clusters of individuals all in poor physical states appeared.

Calhoun referred to such aggregations as "behavioral sinks". They are self-stimulating aggregations of individuals often collectively showing abnormal behaviour patterns. Some individuals become inactive and socially withdrawn. As these conditions progress only limited portions of the environment are utilized but these become excessively polluted.

Thus, the stage is set for both high social mortality and high disease mortality and population often collapses as rapidly as it built up. After the crash if the environment recovers the surviving individuals recuperate and population growth may begin again. The detailed of this theme differ considerably in different studies and different species. In come circumstances, there may be a cessation of production so that no young are born into the crowded population, as occurred in house mouse populations studied by Crowcroft and Row.

In other circumstances reproduction may continue, but there may be massive infant mortality of all young born. In still other cases, crowding may stimulate excessive migration and disposal as in lemmings and these mass migrations are often accompanied or determinated by high mortality. In all cases, a similar syndrome prevails. There is usually a drastic change in social organization and behaviour which most often have a major influence on reproduction and mortality. In domestic rodents these changes characteristically involve a breakdown in social stability and a disruption of normal territorial, reproductive and parental behaviour. Frequently parinatal processes are most dramatically affected, so that this pre-and postnatal survival of the young is reduced.

All population of living organism exist in a network of interactions with other population. Many of these interactions are subtle and complex than direct food relationships. Others are competitive or limiting to the interacting populations. The degree of intimacy in the relationships of the organism inhabiting a particular habitat varies greatly. These association may be either beneficial or harmful to both the patterns or beneficial to one partner and harmful to the other. If the relationship is beneficial either to one or both partners, it is called 'Positive Interaction' or cooperation. This may be interaspecific (Ex. Social life of inspects) or Inter specific (Ex: Commensalism, Mutualism & Symbiosis). The interrelations between individuals that are harmful to one or more of the participants is called Negative interaction. It includes "Parasitism, Predation and Competition".

Commensalims

The term 'Commensalism' was first proposed by Van-Benedan to the animals that share the same food. "Two or more animals that live together and do not enter into any kind of Physiological Union are referred to as `Commensals' and this association is called as `Commensalism! The relation is unequal in that only one individual gets the benefit from the association while the other is neither benefited nor injured. The Chief benefits from this association are (i) Shelter, (ii) Anchorage (iii) Transportation and (iv) The food supply.

Types of commensals:

A commensal which attaches or lives upon the host's body is called as 'ectocommensal'. If a commensal lives within the body of the host in the respiratory or alimentary tract or in any other body cavity which opens to outside, is referred to as endocommensal.

Vorticella, a Protozoan animal is an ecto-commensal which is attached to the legs of the freshwater shrimp (prawn like) by slender stalks. The 'Sucker fish' is another ectocommensal which attaches itself to the body of shark, or sword fish or seaturtle with the help of its sucker. All the ectocommensals are small in size and not burdensome to the host. The harmless protozoans in the intenstinal tract of mammals and micro-organisms in the canal system of sponges are the examples of an ectocommensals. An example of commensalism is the remora-shark relationship where the remora fish (Echenesis naucrates) attaches to the skin of the shark by means of a strange sucker disc and transported widely and rapidly by the shark's motive power. The remora also consumes food remnants cast off from the jaws of the shark. Thus the remora benefits in two ways from this attachment and the shark is relatively unaffected although its swimming speed may be slightly impeded. Many large organisms provide commensal harborage for other organisms externally and internally. Large tropical trees provide habits for numerous commensal plants and animals.

Trunks and branches provide attachment sites for epiphytes such as orchids, and recesses between buttress roots provide attachment harborage for bats tree frogs, lizards, insects and many other organisms. Whales provide attachment sites for barnacles, algae, and other sessile marine forms. These plants and animals are not parasitic, that is, they do not extract nutrients from the habitat provided by the host. Most animals, including man, also contain internal commensals. The human digestive tract is inhabited by several types of bacteria and protozoa such as Entamobea cole, which are neither parasitic not pathogenic, but merely residual in the alimentary canal.

Another common type of commensalism occurs when various species of plants and animals use burrows or nests constructed by another species. Termite nests provide ecologic niches for more than 100 species of other animals including ants, aphids, beetles millipedes and isopod crustaceans. Tube-dwelling annelids of the genus Chaetopterus which inhabit the tidal zone of the seashore provide habitats for small crabs of the genus Polyonyx. The crab benefits by the protection of the tube, and it apparently does no harm to the annelid worm which originally built the tube. Commensalism may be obligatory, in which one organism depends entirely on another for its habitat or it may be facultative in which each organism may be independent but one is enhanced in the presence of the other. An example of obligatory commensalism is the relationship between the algae, Basicladia, and certain freshwater turtles. The algae grows only on the backs of the turtle.

Similarly the bivalve mollusc Ostrea grows almost exclusively on the roots of the red mangrove off the coast of Florida. Attached to the shell of the horse-shoe crab, Limulus, are several species of molluscs, barnacles and annelid worms which again are not parasitic but only residential and some of them live only on the Limulus shell. The oyster crab is a small crustacean which lives only in the mantle cavity is a small crustacean which lives only in the mantle cavity of oysters. These all represent cases in which one species of plant or animals has evolved a dependent relationship does not involve feeding upon the tissues of the host. The possibility exists that this is one evolutionary route toward parasitism and the distinction between obligatory commensalism and parasitism is often difficult to draw.

Facultative commensalism is represented by the relationship between the borrowing owl (Speotyto cuniculario) and the prairie dog (Cynomys sp.) The burrowing owl often nests in prairie dog burrows but is not confined to such burrows. In the north polar regions, the arctic fox Clo pex lagopus) feeds in the winter time on the remains of seals killed by polar bears (Thalarctos maritimus) though it may have other sources of food as well. Probably the majority of commensal relationship are facultative. It is especially interesting to consider animal populations which form facultative commensal relationships with men. Among vertebrates perhaps the best examples are domestic rodents such as house mice and Norway rats which have increased in abundance and distribution by their association with human habitation and agriculture.

The house sparrow, pigeon and Asian house crow and Asian house shrew or musk shrew have also greatly increased their population sizes through human associations. Cities, towns and farms provide increased food supplies and nesting sites of these animals. The rhesus monkey in India is one of the few primates which has become a facultative commensal of man. The rhesus can readily survive in forests without man, but it becomes especially abundant around villages and towns of the Gangetic basin where it could live on agricultural produce.

Some more examples among the inverterbrates are as follows:

(1) The peacrabs are found living in the mouth cavity of bivalves. They steal food collected by the host and do not cause any injury.

(2) A number of micro organism live as commensals in the canal system of sponges.

(3) Polynoid polychaetes are found in the pallial chambers of gastropods and in the ambulacral groves of star fishes.

(4) A host of commensals such as worms, Brittlestars, molluscs, little fishes, prawns-crabs etc. are found among the coral reefs.

(5) Small pelagic fishes live under the umbrellar shelter of large jelly fishes.

(6) Some fishes live in the borrows of worms and crustaceans.

(7) A number of copepeds live within the Ascidiann or Sea anemones.

(8) Barnacles attach on the skin of whales or back of turtles.

In all these associations one of the partners which is generally smaller derives benefits of anchorage, shelter, food and transportation.

Mutualism

Mutualism is an association between two or more animals in which all derive mutual benefit with no harm to any of the partners. "Symbiosis" term also often used to this type of relationship (Properly refers to the intimate associations to two or more dissimilar organisms). Mutualism may be seen between animals and plants and between animals and other animals.

Ist Type : Mutualism between Plants and Animals

This association is important to the reproduction of many kinds of plant species. Plants attract insects by their sweet, nectar or colour. The insects visit plants to get their food. The plants dust them with pollen and the insects carry the pollen from flower to flower. In the distribution of seeds, plants take the help of animals. Plants provide fruits and seeds pass in their hard coat, undigested through the alimentary canal of the animal and are distributed in its droppings. For instance a species of Alacia grows little pearshaped masses of food for the ants, and hollow thrones (Myrmecolphilous stipules) in which the ants take shelter. In turn the ants defend the plant from the attack of other insects.

2nd Type: Mutualism between Animals and other Animals

Birds display this type of relationship. The cow bird and Bison, the Oxpecker and the antelope, crow and the cattle exhibit mutualistic relationship. In all these associations the birds render considerable service to their hosts by ridding them of ticks and external parasites and also serves as sentinels. In turn the birds receive a constant food supply. An interesting example is between a 'Crocodile' and the bird (Pluvianus aegyptius). The bird removes leeches and fragments of food in decaying condition of the crocodile. The crocodile opens its jaws and permits the search then finally the bird receives the food from the crocodile. The common starling bird may also perform the same function from the sheep. The association of ostriches and Zaebras is also said to be of mutual benefit.

Birds and Insects

Birds have also developed relationship with insects for mutual benefit. The horn bill of Africa harbours a number of insects in its nest. The female horn bill will be confined to the nest at the time of egg laying and breeding. The male bird feeds her with fruits. Various ants plants (Lice Aphids) are associated as Commensals. The most common Commensals among marine forms are the Hydroids and Sea anemones found on the shell of Hermit crab. In this association both of them are benefited. The Indian Rock (Mincus inermic services as a moving home to zoophytes small polyps are encrusted on this fish. In this relationship the fishes obtain protection from predators and they pay for it by luring large fishes within the range of tentacles.

The relationship between *Hermit Crab* and *Sea anemone* is very interesting. Here the Crab obtains protection from predatory fish by the shrinking power of sea anemone. The sea anemone also gets an advantage of transportation of feeding grounds and gets food fragments when the crab captures and eats another animal. Sponges growing on the shell of Hermit Crab protect it by producing bad odour. Here the advantages to the sponge is that it is carried to new waters for its own food. Thus, the mutualistic relationships between plants and animals, and animals and other animals prove to be beneficial mutually and reciprocally to the participants.

Like commensalism, mutualism may be obligatory or faculative. An example of obligatory mutualism is provided by the relationship between the flagellate protozoan Trichonympha, and woodeating termites. These species cannot live without each other.

The protozoan lives only in the digestive tract of the termite. The stomach and caecum of the horse contains millions of ciliate protozoa and bacteria which digest cellulose for the horse and provide 20 per cent of its nitrogen requirement per day. These intestinal microorganisms are essential for the normal growth and health of the horse.

An elaborate form of obligatory mutualism occurs between some ants (e.g., Cremastogaster lineolata) and aphids or plant lice (e.g. Aphis caliginosa). The ants maintain the aphids in specially constructed nests, and they feed upon a nutrient liquid secreted by the aphids. The aphids have sometimes been called "cows" because they release the secretion when stroked by the antennae of the ants. Thus the ants receive food from ants. Many other anthropods, including mites, springtails, flies, wasps and beetles, have developed similar complex relationships are based on interspecific communication systems through which the different species jointly utilize food, shelter of other resources.

Many mutualistic relationships are facultative; that is, both populations can exist without the other, but both are favoured when living together. Baboons and impalas, for example, can survive well without each other, but each provides a unique protective warning system for the other when they live in association. Squirrels facilitate the extension and propagation of hickory trees by buying the nuts, but the hickory trees can propogate without squirrels and the squirrels can likewise survive without hickory nuts, if an adequate supply of other food is available.

Man has capitalized on mutualism as a biological principle in the development of agriculture. The domestication of plants and animals originated in facultative commensalism or mutualism. At the down of agriculture, 10,000 years ago, man could live without domesticated plants and animals, but as he developed these relationships then evolved into obligatory commensalism and mutualism. Modern man obviously cannot live without the domesticated plants and animals of agriculture, and many of these domesticated forms cannot live without the special husbandry provided by man.

Symbiosis

The term Symbiosis was first coined by DeBray in 1879. It means "living together." "Symbiosis is an association between two or more organisms that live together in close physiological union for the mutual benefit." Here the term Symbiosis refers to the intimate association of two or more dissimilar organism. One of the partners may be known as Symbiont.

Symbiosis offers from commensalism and mutualism in two important respects.

(1) The relation between two partner is much more intimate in Symbiosis than in other relations.

(2) It is mostly obligatory.

Symbiotic associations are known to occur between plants and other plants, animals and plants; and animals and other organisms.

First Type Symbiotic Association among Plants

This type of associations is demonstrated in (i) Fungi and algue which form lichens; (ii) Fungal mycorrhizae and the roots of many flowering plants and (iii) Nitrogen fixing bacteria and the roots of legumes. In all these associations, there is metabolic interdependence. In the lichens, algae produces food and fungi provide mineral and protection. The intimate association between the roots of flowering plants and mycorrhizae results in the formation of a composite fungus-roots structure which differs from the roots without mycorrhizae. Here the fungi require soluble carbohydrates and substances and they get them from root secretions. Here also the association is intimate the benefits are mutual. The importance of the Nitrogen-fixing bacteria (Rhizobia associated with legumes is well known. Legumes posses on their rootlets, a number of swellings, inhabited by the bacteria. Here, they derive carbohydrates from the host and in turn they fix the atmospheric nitrogen in the roots. This "Nitrogen" is used by the plants.

Second Type: Symbiotic-Associations Between Plants and Animals

The most notable symbiotic association between plants and animals occur between freshwater coelenterate (Chlorohydra Viridissina) and an algae. In this association the animal obtains Carbohydrates and a supply of Oxygen evolved as a result of Photosynthesis by the algae. The large in turn receive protection and Dioxide for photosynthesis released by the animal during respiration. Planeria and an animal algae and a plant symbiotic association is also a good example. Here the symbiotic benefits are same as in coelenterate and algae. Here, the presence of algae is indispensable for the worm and this is obligatory.

Association Between Fungus and Insects

The wood boring beetles have an elaborate apparatus on the head for carrying the Fungus. Others carry Fungus in the midgust to be regurgitated for cultivation. The females of this species have an apparatus connected with evipositor which coat their eggs with fungal spores as they are laid. In all these associations, the Fungus provide food and shelter, and it grows on the excretory material of its host.

Third Type: Symbiotic Association Between Animals (Metazoa) and Microorganisms

Symbiotic associations are also seen between Metazoa and Micro-organisms many Wood-eating-insects especially termites harbour numerous protozoans in their alimentary canal. It has been shown that the gut of termites can live on a nitrogen-free diet so long as protozoa are present. This suggests that they may be capable of nitrogen fixation of the synthesis of protein from carbohydrates. Here the protozoans produce specific enzymes "Cellulose," and "Cellobiase," which are responsible for cellulose digestion. Similar association is seen in the gut of Cockroach, "Cryptocersus." In all these cases the protozoan gets shelter and food.

Mammals and Bacteria

In this association symbolic bacteria are present in the Caecum of rabbits and rumen of cattle. Here, the bacteria utilize the inorganic nitrogen and sulphates for the synthesis of essential amino acids, which can't be synthesized by the animal. In turn to mammals, the Bacteria produce Vitamin B_{12} (in ruminants). In man also some of the vitamins are produced by the intestinal Bacteria. The Symbolic Bacteria are also present in the gut of the South African Honey Guide." Here, the Bacteria feeds on the Wax of honey comb the bacteria digest the Wax and gets shelter and food from the host.

Predation

In this relationship one animal is killed and eaten by another, the predator so predator is a free living animal. It is usually larger than prey. The herbivores that feed on the plant material completely are called as Plant

Predators. There are, many kinds of predation which are not necessarily so selective. The swallow feeding on moths and midges, the top minnow or trout feeding on mosquitoes, and the ant eater feeding on an ant nest is less discriminate in taking individuals.

Certainly, many able and healthy individuals are preyed upon. In cases such as these, which involve large numbers, the prey population often shows great productivity which can accommodate this mass mortality. Many fish, insects, and other Invertebrates have tremendous production of eggs and larvae, but only a small percentage of these serve the onslaughts of predation and other forms of loss. For example, the horse shoe crab or Limulus of estuarine and evastal waters deposits its eggs on sandy beaches wehere they are consumed in great numbers by flocks of laughing gulls and other shore-birds.

Large number of eggs also wash out to sea and are eaten by fish, whereas others are pushed up on the beach and dry out. Despite this fantastic loss of eggs and young larvae, Limulus continues to survive in large populations and has done so for the last 180 million years. It is one of the oldest animals on earth in terms of its evolutionary history, having outlived trilobites and dinosaurs by millions of years, and it is still incredibly abundant. There has been considerable interest in predation as a factor in quantitative population control as well as its role in quality control.

On one hand, sportsmen have wanted to limit predation, being concerned that predators may take excessive numbers of game birds and wildlife. Some sportsmen's organizations have therefore supported predator control campaigns against hawks, owls, foxes, coyotes, wolves, cougar and other predators in the name of conservation.

On the other hand, ecologists have often been interested in promoting predation as a method of biological control of pest animals, particularly insects and rodents, and even as a necessary biological control of desirable animals such as deer, or prevent them from becoming pests.

There are also examples from the entomological literature where predation serves as a significant population control on an animal population. In a classic study of natural population balance in the knapweed galifly (Urophora jaceana), a small dipteran insect, Varley concluded that one of the major controlling factors was the predation of mince upon insect pupae. His data showed that 22 to 43 percent of the fly morality was attributable to this predation.

One study of tsetse flies (Glossing spp.) in Africa, found predatory spiders to be a controlling factor on the abundance and distribution of the flies. Insect predators have also been used in applied biological control of agricultural pests in a few cases. In California, the cottony cushion scale, an hemiptoran insect pest of citrus crops, was effectively controlled by a predatory ladybird bettle imported from Australia. In Hawaii, the sugarcane leaf hopper has been controlled by the capsid bug.

Another important aspect of predation as an influence in population ecology and natural selection is the great variety of adaptations which have arisen in response to it. Protective coloration, warning coloration and mimicry fall into this category. Also, the morphological and behaviour, freezing behaviour and many other specialized patterns have developed in various species as adaptations to reduce mortality through predation.

One particularly interesting system which has been studied in some detail is the intimate relationship between certain moths and the bats which prey upon them. The bats locate the moths by emitting ultrasonic pulses and detecting reflecting echoes flying moths, a natural sonar system.

The moths have evolved the ability to detect these ultrasonic pulses of the bats, and upon perceiving the approach of a feeding bat, they instantly go into complicatd diversionary flight spirals to avoid the bat. The introduction becomes on serial "dog flight" between bat and moth, with the moth trying to avoid the bat. This ability and behaviour has evolved in selective response to predation in the same way that cryptic coloration has evolved. In both case the advantageous mutations have been selected and propagated.

Parasitism

"Parasitism" is the relation between two individuals, wherein the parasite receives benefit at the expense of the host, therefore, parasitism is form of disoperation. In this association, the parasite receives (i) food (ii) shelter (iii) and protection, at the expense of the host. The host does not derive any benefit from the relationship. A parasite does not ordinarily kill its host atleast not until the parasite completes its reproductive cycle.

Classification

Parasites are classified into different categories depending upon their relation with the host.

1. Temporary Parasites

   The animals which spend only a part of their life cycle as parasites are called temporary parasites."

   For Example; The glochidium (is a Larva of freshwater mussel) leads a parasitic life.

   It attaches itself the skin of fish with the help of hooks present on the shell and later it is covered by the skin of fish. It remains burried thus for several weeks and emerges as young animals to lead a free life.

Example 2. In prasitic Hymenoptera, the eggs are deposited into the body of another insect (Caterpillar), where the entire larvae stage is spent the larva feeds on the fat body of the host. Until it attains pupal stage. Finally it emerges as an adult and leads a free living life.

2. Permanent Parasites

Some parasites have continuous association with their hosts, they are called permanent parasites. They spend their entire life as parasites.

For example: Trichinea worm, Malarial Parasite etc.,

Permanent prarasites often found among lower animals. They almost always require a succession of hosts or intermediat host.

3. Facultative Parasites

These are parasites (animals) which are parasitic only when there is a change for parasitic life, leading free life otherwise.

Eg: Meggocs of some flies, Lamprey etc.,

4. Obligatory Parasite

Some species are forced to lead a Parasitic life. If not, they can't live. This type of parasites are called obligatory parasites.

Eg: Filarial Wornm Cestodes etc.,

5. External Parasites

Those which live outside the body of the host are called external parasites. Eg. Flea on birds and mammals, blood sucking lice and fleas, mites, ticks and leeches.

6. Endoparasites

These are parasites which live in the elementary tract body cavities, various organs, blood or other tissue of the host. The association between 'endoparasites' and their hosts is invariably more inntimate than that between extoparasites and hosts.

Tricchom^nad flagellates, Opalnid ciliates Sporozoans, Nematodes, Cestodes and some Copepods are some of the endoparasitic forms.

The endo-parasites are further divided into 2 types based upon their exact location inside the most body.

(i) Intracellular Parasites Ex:- Plasmodium (living within the cell)

(ii) Intercellular Parasites Ex:- Tryponasome (living be tween the cells in the intercellualar spaces).

Evolution of Paraitism

The origin of parasitic life in amimals is associated with attempts to secure food for themselves or their offspring. The ancestors of ecto-parasites were clearly free living forms. A free living organism could accidentally have settled on the body of a larger species where coditions were favourable for survival. For example: The biting lice might have evolved from the insects that live beneath the bark of trees. Endoparasites might have evolved from ectoparasites or directly from free-living ancestors.

Modifications and Adaptations

1. Organs of attachment: The rasping mouth of the Lampern, and the head of tape worm equipped with 'Suckers' and 'Hooks' for attachment to the hosts gutwall are the typical examples. Among parasitic arthropods such as mites other insects claws are well developed used to cling on the external surface of the host.

2. *Body Shape:* The intestinal parasites such as the tape worm and Round worm have elongated bodies. The liver fluke have leaf-like bodies and elongated form of blood flukes are good example of this type. These arrangements are useful to suit the environment in which the parasite lives.

3. *Cuticles:* Ex: Tape worm.

Intestinal parasites face the problem of with-standing the action of the digestive juices of the host. To avoid this these parasities develop thick cuticle, offers resistance against the action of the digestive juices.

4. *Loss of Locomotary Organs:* Ex: The fleas or Lice.

Most of the endoparasites have lost their Locomotary organs, as they are not needed. There is no need in search of food, as they have always plentiful supply of food. So the locomotary organs of the parasites disappear or differentiate into organs of attachment.

The fleas or lice (Parasitic Insects) have lost their wings, which live among the hair and feathers.

5. *Digestive System:*

In the tape worm, which lives in the alimentary canal of the hot, has lost its alimentary canal and absorbs already digested food from the surrounding medium. The mouth parts of the blood sucking insects have developed 'Piercing and Sucking' apparatus.

6. *Loss of Sense Organs:*

As the internal environment of the host is more or less uniform, there is no use of the sense organs among the endoparasites are provided only with tango receptors.

7. Complicated Reproduction: Ex: Flat Worm, Round Worm, etc.

The reproductive function in most parasites has undergone considerable complication. It is the goal of all the functions of the organism. The endoparasites are characterized by enormous fertility. For example (i) the human round worm lay (Ascaris Lumbricoides) eggs at a rate of 1500 a day (ii) the termatodes and cestodes lay eggs continuously (iii) Taenia solium (Tape Worm) lay eggs annually and have been estimated at 80 million (per year).

The most remarkable feature of flat Worm parasities is the presence of 'complicated reproduction'.

The effect of "Parasitism" on the host is usually harmful. The endoprasites causes effects, more marked than those by extoparasites.

The following Pathological effects caused by parasites in their hosts.

(1) The destruction of cells and tissues of the host by the parasite's movement or feeding act.

    Ex: Entamoeba histolytica-eats the tissue cells of colon and red blood ceils of the host, and plasmodium feeds on liver cells and erythrocytes (R.B.C.)

(2) Parasites may cause 'enlargements's and disorders of lymph glands, and liver.

    Ex: Leishmania.

(3) Parasites may cause 'Malaria', (Plasmodium parasite) and 'Trypanosoma gambiense' diseases due to the secretion of toxins (Poisonous).

(4) The effects of endparasites may also be mechanical. Ex: Ascaris, is a good example when these parasites

multiply in greater number they block the intestine, preventing the feree flow of food, which leads to the death of the host.

(5) The most remarkable effects of prasitism on the sexual characters are seen among crustaceans. The presence of sacculina bring spectacular changes in its host (Crab), which may result finally in death (due to the parasitic castration).

Hyperparasites

Sometimes a parasite harbors a number of parasites in its body. They are called 'Hyperparasites'. A typical example is, small mammal ones as 'Mouse' which invariably harbours a no of ectoparasitic fleas, which themselves contain endoparasites (Protozoans) in their alimentary Canal.

Parthenogenic and Non Parthenogenic Parasites

The parasite which do not cause any injury to the host are called 'Non parthenogenic parasite'. The parasite which cause disease is known as "Parthenogenic Parasite". In man's digestive system 'Bacteria' are helpful in synthesizing' Vitamins, used to man. Some Bacteria causes numerous diseases as pneumonia, tuberculosis, diptheria, bubonic plague, cholera, syphilis and wooping cough etc. Viruses causes, small pox, polio, yellow fever etc. Animal-like microbes causes malaria, (Plasmodium) sleeeping sickness (Trypanosona gambiense) and dysentery (Entamoeba histolytica).

Parasitism is virtually universal in all plants and animals in vertebrate animals, endoparasites are found within many of the major organ systems of the body, most commonly in the digestive, circulatory, respiratory and urogenitial systems. Ectoparasites occur on or within the skin and its appendages, such as hair scales. Throughout most of the world, man is beset with many parasitic organisms including intestinal worms (tape-worms, flukes, roundworms, etc.), intestinal protozoa (amoeba, ciliates and flagelates), parasites of the blood (microfilaria, malaria protozoa, etc.), and a variety of extoparasites (Mice, mites, ticks, mosquitoes, etc.).

Many infectious microorganisms including bacteria and viruses are also parasites in the sense that they extract nutrients from the host. Some of these are pathogenic, that is, they impart the health and normal functioning of the host, as for example, the bacterial organisms of the genus shigella which cause bacillary dysentery. Others, such as some intestinal flagellates, are nonpathogenic and do not harm the host.

Many parasitic organisms may be pathogenic in one individual and nonpathogenic in another. For example, most individuals, infected with Entamoeba histolytica in the digestive tract do not have amoebic dysentery, and, in fact, may have no disease symptoms at all, but in other individuals, the same organism may produce severe disease and even death. This is a matter of individual ecology and immunology which is not well understood.

Theoretically, Entamoeba histolytica feeding on tissues of the host is a parasite, whereas the flagellate Trichomores hominis, feeding on digested foods before incorporation into host tissue is not a parasite, nor is Entamoeba coli feeding on undigested particles of food and intestinal bacterias. Neither is there a sharp line of distinction between a parasite and a predator. Houseflies and vampire bats bite and suck blood form livestock, and they might be considered either parasites or predators. Normally, we think of predators as killing their prey in a short time. Man's relation to the cow is predatory when he kills it for beef, but it is parasitic when he milks it for dairy products. There are some relationships, however, in which the distinction is arbitrary, as for example, in the lamprey feeding on a host fish. The lamprey may kill the fish in a few days or a few weeks, depending upon its size in relation to the fish, and may be considered either a predator or a prasite.

Ecologically, one of the most interesting questions about parasitism is the effect of the parasite population upon the host population. The oldest and best adapted parasites have little or no pathogenic effect upon the host. Pinworm infections in children are relatively harmless, though they may cause a minor irritation around the anus.

In West Bengal, studies have shown that over 75 percent of the local population harbors hookworm (Necator sp. and Ancylostoma sp.), but there is no demonstrable pathology or effect on the people. The infectious burden (i.e., the number of parasites per person) seems to be small and insignificant. Most wild animals have parasites are able to maintian excellent health with these parasites. Zebras in Kenya are intensively infected with many internal and external parasites, but they are usually in remarkably robust health.

There are numerous examples where parasitic infections are harmful and debiliting to the host. Certainly, hookworm necator americana in the southern United States, as it occurred 50 years ago, was a debilitating disease, sapping the strength and health of the infected person. Throughout the world, malaria still remains as a major health problem and it causes a great amount of illness and human misery. One of the most important debilitating disease of modern times is schistosomiasis an infection of blood flukes of the genus Shistosoma. This is a waterborne infection and has spread throughout the tropical world with the expansion of irrigated agriculture.

In animal populations, parasites may also weaken infected individuals. In the bighorn sheep of Wyoming and Idaho, lungworm infections are still a major cause of illness and mortality. In dogs and wolves, heartworms reduce the animal's vitality and hunting succss. In meadow vole populations, botfly parasitism causes larger spleen weights in infected mice and reduces their survival when exposed to cold temperatures.

Periodic reductions of field voles in Wales have attributed in some years to heavy infections with tuberculosis among the mice, but in other years similar reductions occurred without the infection. Similarly, reductions in populations of red grouse in Scotland have been attributed in some years to the round worm parasite, Trichastrongylosus pergracilis, but this has not a consistent finding in all year.

Occasional heavy mortality of gray squirrels in Baltimore has been associated with coccidiosis, caused by a protozoan parasite of the digestive tract. In 1971 a virus disease of wildfowl decimated 90 percent of the pheasants of England, in much the same way as myxomatosis reduced rabbit populations of England, and Europe in the 1950's. Thus, parasitism as an interspecific population relationship may be balanced or unbalanced in regard to its effect upon the host. A seriously unbalanced relationship, in which the host dies is, of course, selectively disadvantageous to the parasite as well.

Another outstanding feature of parasitism is the interspecific complexity which was evolved in some species. Many parasites have primary and secondary hosts and, in some cases, tertiary hosts for various stages of the life cycle. The malaria organism Plasmodium falciparum, for example has parts of its life cycle, the sexual phase of reproduction, within the mosquito, and part, the sexual phase of reproduction within the human host. Schistosomes infect man or other vertebrate animals as the primary host for sexual reproduction. The Chinese liver fluke, Clonorchis sinensis, infects man as a primary host, aquatic snails as a secondary host, and fresh-water fish as tertiary hosts. Despite the complexity of this life cycle, it is a very successful and widespread parasite, infecting millions of people throughout the Orient.

Competition

The word "competition" denotes striving for the same thing. At the ecological level competition becomes important when two organisms strive for something that is not in adequate supply for both of them. Thus, plants compete for light and nutrients in a forest, and animals compete for food and shelter when the latter are relatively scarce in terms of the density of animals. If the population consists of only a few scattered individuals, competition will not be a factor of ecological importance.

In the arctic, for example, plants may be so few and scattered due to the severe climate that no competition for light occurs. The result of competition is that both parties (that is, the competition) are hampered in some manner. At the population level this means that density or rate of population energy flow, will be reduced or held in check by the competitive action. There is another type occurs when two organisms interfere with one another while striving for something even though that something is not in short supply. For example, the organisms might secrete substances that interfere with each other, or they might even eat each other. Many ecologists prefer not to include such direct mutual inhibition under the heading of competition. Both intraspecific and interspecfic competition can be very important in determining the kinds and numbers of organisms. Intra-specific competition is an important factor in those populations that tend to be selfregulated in the manner described above. An interesting behaviour pattern, which results in intraspecific competition for space and a rather effective control of population size, is known as territoriality; it is characteristic of many species of birds and some other higher animals.

At the beginning of the breeding season the male of a territorial species of bird will "stake out" a definite area of its habitat (that is, the "territory") and defend it against other males of the same species, with the result that no other ale is allowed to enter the area. Much of the loud bird song we hear in the spring is for the purpose of announcing to other males "ownership" of land, and not for the purpose of wooing the female, as is often supposed.

A male that is successful in holding his land has a high probability of mating and nesting, while a male that is unable to establish such a territory will not breed. Once the pair is formed, the female also joins in the defense of the territory in many species.

The defended territory also serves a positive purpose of insuring the complex business of caring for the nest and young will not be interrupted by the presence of other males and females. Finally, it should be mentioned that the actual defense of the territory does not usually involve much actual fighting or other severe stress. Would be invaders respect the established bird; loud songs or threat displays usually are sufficient to cause the invader to withdraw.

However, if one of the pair should be killed it is very often quickly replaced by a bird form a reservoir of individuals not established. Therefore, since the territorial behaviour pattern helps avoid both overcrowding and undercrowding it can be siad to be regulatory in the sense that such behaviour promotes a sigmoid growth pattern, or at least helps prevent an overshoot of carrying capacity. Plant populations, as well as animal populations, may regulate themselves, so to speak, and avoid overcrowding. As one drives through the desert in the southwestern United States one is impressed with the fact that desert shrubs are widely spaced, often almost uniformly distributed as if planted at regularly spaced intervals. The pattern would seem to be explained by competition for scarce water, which eliminates all but one individual in a given area.

However, evolution of self-regulation of the population has occurred in some species, in that severe competition for water is avoided by the production of leaf or root hormones that inhibit development of other individuals in the neighborhood. That is, chemical substances rleased by decaying leaves that fall to the ground under the shrub, or by the living root in the soil, inhibit or kill any seedlings that may start to sprout. Such a control mechanism would tend to keep plants spaced apart, thereby reducing actual competition for water, which might result in stunting or death of all plants should there be many trying to grow in the same place.

Amensalism and Antibiosis

Amensalism is an interspecific relationship in which one population is inhibited while the other is unaffected. A simple example is the shading out of certain plants under tall trees. The trees reduce the available sunshine at the ground level, and numerous species of plants cannot find adequate light in the shade. Hence, only shade-tolerant plants with lower light requirements can survive as ground cover in the forest. Antibiosis is a specific type of amensalism in which one organism produces a metabolite that is toxic to other organisms. The best known example is the mold, Penicillium, which produces an antibiotic substances causing the death of many bacteria. This was observed and identified in the classic work of Florey and Fleming in England in the late 1920's and early 1980's which led to the development of antibiotics in clinical medicine. Penicillin, strepotomycin, aureomycin and other antibiotics used against disease organisms all represent a similar ecologic principle, and one which is widespread in nature. Many species of fungi and lichens produce metabolic substances which inhibit bacterial growth. The green algae Chlorella produces substances which inhibit the growth of diatoms of the genus Nitzschia, and conversely, Nitzschia produces a substance which inhibits Chlorella. Many plants also produce substances which are toxic or inhibitory to animals. Old cells of Chlorella produce a substance which inhibits the feeding of Daphnia. Algal blooms of some blue-green and red algae produce chemicals toxic to fish. The well known Red tide of southern coastal waters, produced by the flagellate Gymnodinium brevis, can result in massive fish kills.

Land plants also produce inhibitory substances. The roots of some species of trees produce substances which inhibits the growth of other trees, both of their own species and different species. This may result in a characteristic spacing pattern of trees within a forest, dependent not so much upon competition for sunlight at the conopy level, but upon competition for space, water or nutrients at the subterranean root level.

Many plants produce substances toxic to animals which touch them or eat them. In Nevada and Utah the desert bush Halogeton glameratus produces oxalic acid which is poisonous to sheep. Poisonous toadstools to and tubers have been well known throughout the history of man. In South America and Africa, many native people harvest the tubers of the poisonous manioc plant, but they place it in a series of soaking vats to leach out the toxic chemicals before consuming the starchy tissue of the roots. A small African monkey, the talapoin monkey (Cercopithecus talapoin), has learned to steal the manioc from the soaking vats at just the proper state. Many of these substances are in a class of compounds known as ectorines of environmental harmones. Ectorines are substances produced by the organism which affect other organisms of the same or different species. Not all ectocrines are toxic or inhibitory, of course; many are stimulatory or beneficial. But collectively, they represent chemical messengers by which individuals and species are interrelated.

Aggregations

All organisms react on their habitat in one way or another, and when they occur in numbers these reactions produce a conspicuous effect. Water conditioning occurs hence physical or chemical changes occur as the result of organisms living in it. Compared with unconditioned water, these changes may have either a harmful or a beneficial effect on organisms introduced into the water after the original organisms have been removed. Water is said to be homotypically conditioned when the changes were previously produced by individuals of the same species as being studied and heterotypically conditioned when the changes were produced by a different species.

Experimental studies have demonstrated that goldfish grow faster in water that has been homotypicaly conditioned for 24 hours than in unconditioned water. Both fish and amphibian larvae also do better in water conditioned by the presence of mollusknes than in unconditioned water.

The marine flatworm Procerodes wheatland will survive much longer when transferred to fresh water conditioned by the presence of either live or dead individuals of the same species or by freshwater species of flatworms than they do in unconditioned fresh water. The longer survival in toxic solutions, faster growth, and greater reproduction of protozoans, snails, flatworms, cladcerans, amphibian larvae, and fish occurring in aggregations rather than as isolated individuals is attributable to water conditioning. Benefits derived from aggregating are shown in other ways by terrestrial animals. In honeybees, when hive temperatures drop below 14°C (57°F) during the winter, they form clusters and maintain a mass temperature several degrees above outside temperatures. This is brought about by increased metabolic oxidation of honey in their bodies and by increased muscular activity.

Furthermore, the compact cluster presents a surface area for heat loss that is less than the total surface area of the individuals separately. When there is danger of overheating, the bees in the hive spread out on the combs and fan with their wings to create a circulation of air. They will also carry water into the hive and place small quantities both outside and inside the comb cells. The forced air circulation evaporates the water and cools the hive. Bees also cool themselves by constantly moving their tongues in and out of their mouths, exposing to evaporation the moisture that is present on them as a thin film temperature regulation is less well developed in other social Hymenoptera.

Allelochemistry

Included here are the coactions whereby chemicals secreted by one organism affects the growth, health, or behaviour of other organisms. Allelopathy is produced in plants when toxins are liberated that inhibit seedling growth in the vicinity. This may affect succession of plant species, especially important in the early stages. Some pioneer species in the abandoned field sere produce substances inhibitory to nitorgen-fixing and nitrifying bacteria. This retards invasion of other species that require higher nitrogen concentration in the soil. Volatile inhibitors are generally more prevalent than water-soluble ones, and relatively more prevalent in arid than humid climates. Antibiotics produced by bacteria, fungi, actinomycetes, and lichens are widespread in nature and may be one of the reasons why bacteria pathogenic to man cannot mutiply well in soils. A number of antibiotics, such as penicillin, have been used extensively in human medicine. The use of allelochemic coactions agriculture has possibilities but has yet to be exploited. Among animals, overcrowding of tadpoles in culture dishes in associated with the occurrence of peculiar round vacuolated cells in the intestinal tract and feces that appear responsible for curtailment of further growth.

Under laboratory conditions, killer stocks of Paramecium aurelia produce a toxin, paramecin, at the rate of one unit-particle per animals per 5 hours. One unit-particle is enough to kill one individual of sensitive stock of the same species as well as being lethal to other species of Paramecium. Conditioning that becomes unfavourable \homotypically may sometimes be favorable, or at least tolerable, heterotypically. Thus in protozoan infusions there is a microsere of one species succeeding another. Allechemic effects are of great variety in both plants and animals: repellants, escape substances, suppressants, venoms, inductants, attractants, signals, stimulants, autotoxins, antoinhibitors, and so on. Pheromones chemical messages between members of a species especially important in reproductive behaviour, social regulation and recognition, alarm and defense, territory and trail marking, food location, and so on.

Protocooperation

In the positive relationship between populations, proto-cooperations is a short step ahead commensalism, to cooperation, both populations gain by the association or interaction of some kind. A considerable work on protocooperation was one by Alee. In nature, crabs and coelenterates (sea anemone) are often found associated with each other. This association is of mutual benefit to both. The coelenterates grow on the back of the crabs, or sometimes planted there by the crabs. The coelenterates provide protection to the crabs.

There exists complex relationships between different populations. Some of these relashionships play a vital role in the regulation of populations. Successful parasitism represents something of compromise between two populations. Predator-prey system is the product of long evolutionary process. A close relationship exists prey, but it is not always the case. The inter-specific competition is a basis to the theory of natural selection and evolution of species.

9 ENERGY RESOURCES

The world's leading experts now agree that burning fossil fuels— coal, oil and gas — is slowly but relentlessly upsetting our climate, releasing carbon dioxide which traps heat closed to the Earth— the so-called 'greenhouse effect'. Since 1850 the concentration of carbon dioxide in the atmosphere has increased by a quarter. An expert panel has reported to the United Nations that the Earth's average temperature may already be rising. In the past ten years we have had seven of the eight hottest years ever recorded. A rise of even a degree or two in average temperature worldwide may be catastrophic. As oceans warm they expand. Together with the water from melting ice, this will cause sea levels to rise resulting in the flodding of coasts and islands, even some town and cities. Croplands may suffer from droughts. Deserts may spread. Animals and plants may not survive the change in their local climate. Weather may become even more unpredictable, with violent winds, storms and hurricanes.

The poorest countries, least to blame for the problem, are the least able to adapt and will be hardest hit.The average North American now uses 16 times more energy than the average African. The gap between the industrial North and the developing South, between the rich in the cities and the poor in the countryside, will widen even further. The South's crushing burden of debt, much of it for importing fuel, will worsen, leaving most of the world's people even farther behind, aggravating social and political tensions. Nearly four times more energy is now used worldwide compared to 40 years ago. Official forecasts all assume that we will double or even triple the amount of coal, oil and gas we use worldwide, within the next 40 years. They also assume major expansion of nuclear power, despite staggering costs, unsolved problems of safety and waste disposal, and unbreakable links with nuclear weapons.

The cost of doubling or tripling suppllies of fossil fuels and nuclear energy will be awesome. Think of one new coal-fired and electricity plant every two days, one new giant oilfield every month or two, one or two new nuclear power plants every two months — for 40 years. Think, too, of the environmental consequences, both local and global. The price of fossil fuels and nuclear power does not include the cost of their damaging side-effects. We are paying the cost, and so is the planet we are living on. Some insist we must wait for absolute proof of global warming. Unfortunately, by that time it will be too late. The risk is too large for us to take a chance. Delay wastes not only valuable time but valuable opportunities. Furthermore we already know what to do.

The key is a change in perception, one that has been advocated by some energy experts for at least 20 years. Nobody actually wants coal, oil, natural gas, or electricity. We want comfortable buildings, light and heat, cooked food, and the ability to exert forces, lift weights and move around easily. We don't want energy— we want the services that energy delivers. By improving our buildings, lighting, applicances, industrial processes and vehicles we can get better services using less fuel and electricity, with lower costs and less damage to our environment. Furthermore we can get the fuels and electricity we need not from fossil fuels and nuclear power, but from the natural energy flows around us —

so-called 'renewable' energy. Energy security does not mean a sure supply of coal, oil, gas and electricity; it means a sure supply of energy services. We already know that we can cut down on fossil fuels and nuclear power, and change over to clean, renewable energy. We already know a lot about the technologies for renewable energy and about how to use them with high efficiency. However, many major companies and entire countries make their money by selling traditional fuels and electricity and these habits are hard to change. The problem of the changeover is neither technical nor economic, but political.

Within the next 20 years-within out lifetimes— we will know if we have broken our addiction to fossil fuels and nuclear power. Will we make the Earth unfit for our children to live on? Or will we make the change to efficiency and renewables? We can do it. But time is running out. Energy is the basic need to national economy. Electrical energy is primarily derived from coal, water and oil in India. None of these energy sources are free from causing some impact on environment. Sustainable development aims at identifying such impacts and taking precautionary steps to minimize and intimate them in order to safeguard the environment. Problems of coal production have been diverse right from receive, mining and transportation etc., to political and labour related issues. Oil situation after the gulf war is no less critical and has already created tremendous shares on the country's balance of payment situation. It is evident that conventional forms of energy sources are getting depleted and becoming costly.

India is the seventh largest country in the world (areas 3.28 millions sq km) with a population of 849.93 million but the per capita energy consumption is only 1/8th of the world average. 75 per cent of the population in India lives in rural areas but the energy consumption in rural areas is 1/10th of the commercial energy consumption in urban areas 74 per cent of villages are electrified but only 15 per cent of the rural households are using electricity. However, consumption of electricity in this country has been growing at the rate of 10 per cent per annum compounded over the last 30 years. The increase in the installed generation capacity has not matched the demand rate. In such difficult situations it is essential to tap other sources of energy.

(1) The major thrust of the energy policy is increase in commercial energy supply and

(2) improvement in the standard of living of people particularly in rural areas.

The Department of Non-conventional Energy Sources (DNES) was set up in 1982 which formulates policies and programmes for the research and development activities in New and Renewable Sources of energy and their implementation. The seventh plan had an ambitious target to deliver energy through these sources equivalent to over 40 million tonnes of coal/year.

These alternative sources of energy are discussed here as follows: (1) Solar Energy (2) Biomass (3) Biogas (4) Wind (5) Wave (6) Tide (7) Ocean Thermal Conversion Energy (8) Geo Thermal (9) Nuclear Fission. While all these alternative sources of power generation from the non-conventional energy sources have been making their presence felt in one way or other, they too require huge capital investment and organization of communities through mass movement to make the system viable.

The renewable sources of energy or the alternative form of energy which can save us from the energy crisis and become the major sources of energy in future include: solar energy; waste biomass energy; Sea-wave energy, Tidal energy, Geothermal energy, wind energy and fusion energy. And of course, the 'Energy conservation' is also very important. Let us now discuss all these sources of energy in detail.

1. *Solar Energy*. The sun is almost inexhaustible source of energy. Sun's heat energy is being used directly in solar cookers to cook food or in solar heaters to heat water. The sun's energy can be concentrated by big concave reflectors and used to boil water to obtain steam and produce electricty. Solar cells can convert the sun's energy directly into electricity. The biggest advantage of solar energy is that it does not bring about any pollution.

2. *Waste Bio-Mass Energy.* The waste bio-mass such as cattle dung, plant residues (crop residues) and sewage, etc., on fermentation in the absence of air gives rise to a combustible gas called bio-gas. This can be developed as a good source of energy in rural areas (village areas).

3. *Sea-wave Energy.* The waves on the surface of sea move very rapidly and have a lot of kinetic energy. The energy of moving sea-waves is now being used to generate electricity. Specially shaped devices have been designed which move as waves in the sea pass them. This movement can be used to drive dynamos to generate electrictity.

4. *Tidal Energy.* The enormous movement of water round the world's coasts between high and low tides provides a very large source of energy. This energy can be tapped by constructing a tidal barrage. The difference in the depth of water on either side of the barrage is being used to drive turbines and produce electricity.

5. *Wind Energy.* The traditional wind-mill is well known but it is not used these days. Newer type of wind-mills are being developed which could be a good source of energy in exposed, windy places.

6. *Nuclear Fusion Energy.* All the existing nuclear power stations are based on the nuclear fission principle, the source of energy being the fission of radioactive substances such as uranium-235. But the supply of uranium-235 is limited and hence cannot last for a long time to come.

Another type of nuclear reaction which can become an extremely good source of energy for future is the fusion together of heavy hydrogen nuclei to form helium nucleus with a tremendous release of energy. This nculear fusion reaction takes place at an uncontrolled rate in a hydrogen bomb explosion. Scientists are trying to control the nuclear fusion reactions in order to produce a nuclear fusion power station. If they are successful, then we would have a virtually unlimited source of energy because the oceans contain enough of heavy hydrogen (in the form of heavy water) to last for many thousand million years in the nuclear fusion reactors.

8. *Energy Conservation.* It is a common saying that "Energy Saved Produced". So, we should try to prevent the energy from being wasted uselessly. This can be done everywhere, in home, in school, in transport and in industry. In many homes; the lights and fans are kept switched on even when not needed; radio and T.V. are kept on unnecessarily; and water gets heated to high temperature by sheer negligence and then allowed to cool for taking bath in winter. All these are highly undesirable practices which waste the precious electrical energy. So whenever you see a fan or light on unnecessarily in your home or school, please switch it off at once. We

should take a great care in using the fossil fuels like coal, petroleum products (like petrol, kerosene and diesel), and natural gas. These should be conserved as far as possible and used when no alternative source of energy is available.

Solar Energy

It is the source of energy driven by thermonuclear fusion reactions deep in sun's core. The earth receives 170 trillion kilowatts of radiation from the sun— Eight days of sunshine is equal to all the available sources of energy in the world put together. A world population of 10 billion with a total power need/person of 10 kw would require 10"kw of energy. It is thus apparent that if irradiance of only 1 per cent of the earths surface could be converted into useful energy with 10 per cent efficiency, solar energy could fulfill needs of entire population.

— *Devices Based on Solar Energy:* The various devices which are being used to convert the solar energy into usable form are: solar cooker, solar heater, solar power plants, solar furnaces and solar cells. We shall discuss all these appliances one by one.

— *Solar Cooker:* A solar cooker is a device which is used to cook food by utilizing the energy radiated by the sun. A solar cooker consists of an insulated metal box or wooden box B, which is painted all black from inside. There is a thick glass sheet Gas a cover over the box, and a plane mirror reflector R is also attached to the box as snown in. The food to be cooked is placed in a metal container C, and the container is kept in the box and covered with glass sheet. The food container is painted black from outside. In order to cook the food, the solar cooker is kept in the sunlight and its reflector R is adjusted in such a way that a strong beam of sunlight falls over the cooker top. We shall now describe how this solar cooker works.

When the sun rays fall on the reflector R, the reflector sends them to the top of solar cooker in the form of a strong beam of sunlight. The sunrays pass through the glass sheet cover and get absorbed by the black surface in the box. Now, about one-third of sun-rays consist of infra-red rays which have a heating effect, and once the infra-red rays enter the box, then the glass sheet does not allow them to go back.

In this way, more and more of infra-red rays (or heating rays) of the sun get trapped in the black box, due the which the temperature in the box rises to about 100° C. This heat cooks the food kept in the black containers. The use of a black surface and black containers inside the solar cooker box is due to the fact that a black surface absorbs more heat than any other colour.

— *Solar Heater:* A solar heater is a device to supply hot water by using solar energy. A solar water heater can be installed on top of flat-roofed buildings and provide domestic hot water throughout the year in a climate like India.

It consists of an insulated water storage tank with an inlet for cold water and an outlet for hot water. The cold water is passed through an iron pipe fitted with a collector. The water absorbs heat from the collector, becomes, hot, moves upwards into the upper portion of storage tank and then can be taken out through an outlet for use. Such solar water heaters are still not widely used because of the high cost of initial installation and because the conventional sources of energy are still readily available. The day is not too far when we will have to switch over to solar heaters to conserve fossil fuels.

— *Solar Still:* This device of obtaining fresh water from sea-water is similar to roof-type solar systems. It is made of a flat rectangular tube about 3 meter long, 2 meter wide and 0.20 meter high. Its upper part is slanting towards the base and also has a U-shaped construction to let the fresh water flow down into a container. It has a blackened surface to absorb more sunlight and is covered with a transparent cover made of glass or plastic film which allows the sunlight to pass through it.

The sun's heat evaporates the water in the tub, leaving behind a residue of salt. The vapour from evaporated water condenses on the surface of the cover and trickles down into containers which thus collects fresh water. These solar stills are being given free to the villages in the areas where there is acute problem of shortage of drinking water.

— *Solar Furnace:* It is an arrangement of mirrors to focus sunlight from a wide area into a small volume. The radius of the earth's orbit is over 200 times the radius of the sun, so if the sunlight on a 2-metre square is concentrated onto 1 square centimetre there is a change of reaching the temperature of the sun's surface.

In fact remarkably high temperatures, in the thousands of degrees, can be reached and sustained with solar furnaces, and they have been valuable research tools for studying the high-temperature properties of materials. Also, such a furnace can be used to raise steam to generate electricity, and small-scale power plant to do this have been built. One of the largest solar furnaces is located at Odeillo in France where temperatures of 4000 C can be reached in a 50 square centimeter hot spot. Parabolic reflectors and other types of mirrors have been used to focus sun's rays on the oven fixed in a ten-storey concrete building.

— *Solar Evaporation:* It is a natural process in which water absorbs sunlight and changes into water-vapour. This is an important industrial process to produce salt from sea-water in many developing countries like India. It is a natural process in which shallow ponds are filled with brine (sea-water). Water allowed to evaporate and crystallized salt collected. About 1 million ton of salt is produced annually using this method.

— *Desalination:* Desalination is the production of pure water from salty or brackish water. Desalination is important for the supply of drinking water or irrigation water in many parts of the world. Desalination on a large scale requires a high energy input, and combined systems have been built using nuclear power to give electricity and also to desalinate sea water. On a small scale, solar energy can readily be used in a 'solar still' solar radiation passes through the transparent plastic cover and is then absorbed by the salt water inside or by the black lining. There is a rise in temperature and humidity within the container, and vapour is driven off from the salt water to condense on the underside of the cover.

— *Solar Distillation:* Solar distillation is a process in which the sun's heat is used to evaporate sea-water in order to produce sodium chloride, other salts and potable water. This is perhaps one of the oldest uses for which human beings have used solar energy. The roof type solar system was used by people in Chile in 1872 and the modern solar distillation plants have the same mechanism.

— *Roof-type Solar System:* Sea-water is put into a shallow receptacle made of concrete and whose inside is blackened with some black material like coke or graphite. The receptable is covered with transparent glass covers.

The water evaporates by absorbing heat of the sunlight and water vapour then condenses on the inside surface of glass covers, drips down and is collected. This water can be used for drinking purposes and also in place of distilled water in labs.

— *Solar Cells:* These are devices which convert the sunlight energy directly into electrical energy or electricity. Solar cells are made from semiconductor elements like silicon and germanium. Solar cells are used to generate electricity needed in man made space-satellites, and to provide electricity in far-off places. For example, in some of village of our country, solar cells are being used to provide street lights at night. A large group of solar cells fixed on the top of a pole, produces and stores electricity during the day when sun shines. And this stored electrical energy is used for lighting at night. It has been estimated that if all the desert areas of Rajasthan (which receive so much of sun-shine) are covered with solar cells the electricity thus generated can meet the energy needs of the whole country. But at present, solar cells are quite expensive. It shoul be noted that these days, solar cells are also used to run watches and calculators.

Until a few years ago any proposals regarding the use of solar energy for electricity production were received with utmost skepticism. This attitude is now beginning to change. Of all forms of energy solar radiation is the one available to maintain the greatest abundance. In the United States and certain other countries such as Japan and Australia,the first steps have already been taken to develop big *central solar power plants.* Owing to the large area of land required and the increase in efficiency derived from solar radiation, such solar power stations should be installed in hot desert or semidesert areas. Another uncomplicated and dependable method of converting solar radiation into electric current has been employed successfully in the supply of energy in spaceships: These are *solar cells or semiconductor chips* which convert the impinging solar radiation directly into electric current. The current flows directly out of the relatively simply constructed semiconductor chips. The reliability of operation is extraordinarily high. Even under severe space conditions a maintenance free life span of 10 or more year was achieved. The only disadvantage of solar cells is their high price.

Solar cells: Solar cells are made semiconductor elements like *silicon* and *germanium.* A large group of solar cells fixed on the top of a pole, produces and stores electricity during the day when sun shines and this stored electrical energy is used for lighting at night. It has been estimated that if all the desert areas of Rajasthan (which receive so much of sunshine) are covered with solar cells the electricity thus generated can meet the energy needs of the whole country. These days solar cells are also used to run watches and calculators. India gets plentiful sunshine for at least 250-300 days/year. In South and Central India the annual average daily incidence of solar energy is 4.5 kwh/m^2 with the maximum exceeding 7.0 kwh/m^2. Therefore solar energy offers an attractive solution to some of the energy problems in India.

Utilization of Solar Energy for Various Purposes

(1) National Physical Laboratory, Delhi has developed a solar water pump at 1kw rating called Abhimanyu.

(2) A solar photovoltaic water pumping system has been developed by Central Electronics Ltd., Sahibabad (U.P.). In this system the sunlight is directly converted into electricity and the energy is used to run an electric motor pumpset.

(3) The Government of India has introduced box type solar cookers with heavy subsidies.

(4) An improvement over the box like look has been devised by REC, Kurukshetra. This is a family size cooker suitable for cooking conventional foods at convenient time for breakfast, lunch and dinner inside the kitchen. It would just replace a bottled gas cooking range and would fit permanently.

(5) Solar refrigerators and solar milk cookers have been designed by the REC, Kurukshetra.

(6) Solar photovoltaic (SPV) technology enables direct conversion of sunlight into electricity thus provide instant electricity in a non-polluting manner— in remote and unelectrified areas for — pumping water for irrigation and drinking, street light, domestic lighting, community TV, wrist watches, calculators railways and road signals, remote microwave communication relays and microwave repeater stations. A number of experimental SPV plants having a range of A1-2.5 kmp have been installed in various parts of the country.

A solar energy centre has been set up at Gurgaon Haryana. A central photovoltaic power system has been installed at village Acchija in Ghaziabad district to provide power for street lights, community TV and radio sets. In September 1992, the Prime Minister of India dedicated to Nation *the amorphons silicon solar cell plant* set up at Gurgaon by BHEL. The develpment of solar energy was taken up as a technology mission is 1986 and the country has made advances in the field by developing amorphons silicon cells. It is proposed to electrify 1 lakh villages through solar energy during 8th plan.

Advantages of Solar Energy: An important advantage of solar energy is that it does not cause any pollution. Moreover, it is available in abundance in a hot country like ours and free of cost. An important limitation of solar energy is that it is not available in sufficient amounts all the time. The supply of solar energy gets reduced during rainy season and during winter months.

Uses of Solar Energy: The important uses of solar energy are given below:

1. Solar energy is used for cooking food. This is done by using solar cookers.
2. Solar energy is used for heating water supply in big building like hotels. This is done by using solar heaters.
3. Solar energy is used for producing electricity needed in space satellites, and to run watches and calculators. This is done by using solar cells.
4. Solar energy is used for the production of large scale electricity in solar power plants.
5. Solar energy is used to melt and forge metals in solar furnaces.

Boimass and Biogas

A variety of biological systems can be used for the production or conversion of energy other than as food. They may be classifed into

(a) *Photosynthesis* by which solar energy is converted into carbohydrates such as cellulose or sugar

(b) Anaerobic fermentation by which plant Biomass may be converted by microorganisms into alcohol and

(c) *Enzymatic breakdown* or digestion of biomass with the production of methane gas.

Biomass: It is the waste material of living objects. It includes cattle dung, wood, sewage, agricultural wastes or crop residues like bagasse (bagasse is the remaining part of sugar cane from which the juice has been extracted) or rice husk.

Electricity from Biomass: Biomass has been used as fuel in traditional homes in the form of cowdung cakes or fuel wood. But now this energy source has attracted the attention of energy technologists. Biomass generation of electricity can solve both the problems of power and unemployment in rural areas. It is said that taluks or tehsils are more suited to run and maintain energy production system on a long term basis. Besides, the solution to a taluks power needs (estimated to be 10-20 MW) means taking care of the energy needs of a large number of villages which constitute it. A recent study has shown that for a 10 MW biomass based power plant, about 83,000 tonne of dry biomass/yr is required. This can either be provided pin crop residues or captive energy plantations.

Indian Scene: India produces about 300 million tonne of crop residues (including rice husk) every year which theoretically can generate about 40,000 MW of electricity. This is nearly equivalent to the total electricity generating capacity to be added during Eighth Plan. In a majority of cases where sufficient crop residues are not available, the only viable option would be a mix of crop residues and plantation wood. The concept of energy plantations with fast growing species of trees like *Lencaena Prosopis, Eucalyptus* and *Acacia* has been mooted precisely for this reason. These yield between 25-30 tonne of dry biomass/hectare-yr in good soils with irrigation and about 10-15 tonne/ha-yr in poor soils with good irrigation. These yields are for a five year crop cycle. Hence for a 10 MW power station, a maximum of 5,500 hac will have to be brought under energy plantations, if it is to run only on plantation wood. Already in Maharashtra, Gujarat, Karnataka and Tamil Nadu large areas have been planted with trees both for timber and fuel wood.

Biomass Power Plants: A 10 MW power plant can give direct employment to about 1,000 people year round for seedling preparation, transplanting, and harvesting of these plantations and for running the power station. These plantations can also provide environmentally beneficial green cover. The infrastructure required for these power plants will be similar to that of sugar factories. The technology for such biomass based power plants is a proven one.

Biogas: Biogas is not a single gas but is a mixture of gases such as methane, carbon dioxide, hydrogen and hydrogen sulphide. Biogas contains 65 per cent of methane gas which is a very good fuel. Biogas is obtained by the anaerobic fermentation of animal wastes like animal dung (or plant wastes) in the presence of water. This fermentation is carried out by anaerobic microorganisms called anaerobic bacteria in the presence of water but in the absence of oxygen. The anaerobic bacteria degrades the carbon compounds present in the animal wastes or plant wastes to yield methane gas.

Biogas plants in India: Two types of biogas plants are used in our country to produce biogas. In these plants cattle dung or gobar is used so they are called gobar gas plants in our villages. The two types of biogas plant being used are

(1) Fixed dome type biogas plant

(2) Floating gas holder type biogas plant

Fixed Dome Type Biogas Plant: Fixed dome type of biogas plant is having a well shaped under groud tank called digester which is made of bricks and a high dome shaped roof D, also made of cement and bricks. The dome of the digester tank acts as a gas holder or storage tank for the biogas. There is a gas outlet S at the top of the dome having a valve V. On the left side of the digester tank is a sloping inlet chamber I and on the right side is a rectangular outlet chamber, both are made of bricks. The inlet chamber is connected to a mixing tank M while the outlet chamber is connected to the overflow tank F.

Cattle dung water are mixed in equal proportions in the mixing tank M to prepare the slurry. This slurry of dung and water get fed into the digester tank T through the inlet chamber I. The digester gets filled with dung slurry upto the cylindrical level, the dome being left free for the collection of biogas. It takes about 50 to 60 days for the new gas plant to become operative.

During this period, the cattle dung undergoes fermentation by anaerobic bacteria in the presence of water with the gradual evolution of biogas. This biogas starts collecting in the dome. As more and more biogas gets collected in the dome it exerts pressure on the slurry in the digester tank and forces the spent slurry to go into overflow tank F through the outlet chamber. This spent dung slurry left after the withdrawal of biogas is rich in nitrogen and phosphorus compounds and hence forms a good *manure*. Once the gas plant becomes operative, more fresh dung slurry is added to the digester tank regularly and gives rise to the continuous gas production. This biogas is also used to run engines and for street lightening.

Floating Gas Holder Type Biogas Plant: It differs from the fixed dome type gobar gas plant in the construction of the gas holder. The dome in it is mobile. As more and more biogas is produced the dome slides up. The floating gas holder type biogas plant is more expensive because it uses a gas holder drum made of steel. Moreover, defects in the welding of gas holder and corrosion of steel can lead to leakage of biogas.

Uses of Biogas

1. Biogas is used for cooking food. When biogas is burned in a gas stove, it produces a lot of heat. This heat finds use for cooking food and for other domestic heating purposes. Biogas is a good domestic fuel because of the following reasons:

   (a) Biogas burns without smoke and hence does not causes air pollution.

   (b) Biogas is having a high calorific value. That is, biogas produces a large amount of heat per unit mass.

   (c) Biogas is a very clean fuel and convenient to use.

   (d) There is no storage problem for biogas as it is supplied by pipes directly from the gas plant.

   (e) Biogas is cheaper than most common fuels.

2. Biogas finds use as fuel to run engines. At many places, the engines of water pumping sets used for irrigation are run on bio-gas, instead of diesel.
3. Bio-gas also find use for street-lighting.

Wind Energy

Moving air is called wind. The wind has energy, the energy possessed by wind is due to its high speed. Wind energy is fast emerging as the most cost effective source of power as it combines the abundance of a natural element with modern technology. Since the energy crisis of 1973-74, interest in wind power has increased throughout the world. The great advantage of wind-driven power stations their environmental harmony. A wind driven power station consumes no raw materials nor does it give off waste gases or other waste materials. Wind-driven power stations would be particularly suitable for use in developing countries as a cheap source of energy, technologically simple to manage and presenting no problems for the environment.

(i) In sail boats for transport purposes, and

(ii) In wind-mills to pump out water from earth, and grind grain to get flour. These days, however, the energy of wind is also being used to produce electricity.

Utilization of Wind Energy in India

In India, efforts to use the latent power in wind began in 1985, when the joint sector Gujarat Wind Farms Ltd. was set up. The company established a wind mill in Mandvi with a total capacity of 1.6 MW. Later the Gujarat Development Agency formed a far more wind forms on its own. In the state of Tamil Nadu the state Government has decided to set up 100 mw of capacity during the eighth plan. The growing interest in wind power technology can be attributed not only to its effectiveness but also to other attractive features like modularity, short project gestation and non polluting nature of the technology. In India, the exercise to harness wind energy includes wind pumps, wind battery charges, stand alone wind electric generators and grid connected wind farms. The Department of Non-conventional Energy Sources (NES) in association with state agencies has been responsible for creating and sustaining interest in the field. But wind power generation received its first impetus during seventh plan when a reassessment of wind potential in the country revealed the possibility of generating 20,000 mw of power.

Uses of Wind-Energy

1. The energy of wind is used to propel the sail-boats in rivers and sea to transport men and materials from one place to another.

2. The energy of wind is used through wind-mills to run pumps to draw water from the ground.
3. The energy of wind is used to run flour-mills to grinds the grains like wheat and corn into flour.
4. These days, wind energy is being used to generate electricity. This is done as follows. The wind rotates the wind-mill blades. When the blades rotate, they turn the coil of an electric generator to produce electricity.

Limitations of Wind Energy

An important limitation of wind energy (or wind power) is that it is not available all the time, and at all the places, to do the various types of work. This is because there is no place in the world where the wind blows all the time. Sometimes we have a lot of wind whereas at other times there is no wind at all. So, without wind, the sail-boats stop sailing and the wind-mills stop turning. There is no surety that we will get wind energy when we meet it most. Despite these limitations, an important advantage of using wind energy is that its use does not cause any pollution. And it is available free of cost! Before we end this discussion, it will be good to list the various uses of the wind energy once again.

Wind Energy in other Countries: This U.S. continues to be a leading use of wind power accounting for about 80 to 90 per cent of the existing installations, the U.S. alone accounts for a capacity of 1400 mw followed by Denmark with 202 mw, other European countries 25 mw and India 32 mw according to figures available upto 1990.

Wind Power Capacity in India: The current estimate for wind power generation in the country in terms of potential is placed around 20,000 mw. Of this, Tamil Nadu alone is expected to account 5,000 mw. However experts say that it will only be possible to generate an estimated 3,000 mw of wind energy by the turn of the century. Wind power capacity of 35 mw is said to have been commissioned by the end of the seventh plan. This includes demonstration project covering a capacity of 31 mv is seven states located at

(1) Mandvi and Lamba in Gujarat

(2) Tuticorin, Kayathan and Muppandal in T.N.

(3) Puri in Orissa

(4) Deogarh in Maharashtra

(5) Tirumala in A.P.

(6) Kheda in M.P. and

(7) Talacauvery in Karnataka.

Wind Movements in the Country: Wind energy utilization very much depend on wind movements at a given place and at a particular time. According to the Tamil Nadu Electricity Board (TNEB) winds in the country are influenced by the South-west monsoons which blow amend May-June. The weaker North-east winter monsoon which starts in October bring in cool, dry air which move towards the ocean. During the period March-August, the winds are uniformly strong over the entire Indian peninsula except of the course the eastern coast. The months of May, June and July account for nearly half the annual energy created. In July-August, wind speeds are high throughout the country being 20-30 kmph in coastal Gujarat; 15-25 kmph over western India, south Tamil Nadu and east coast. In September the winds begin weakening over the country with only coastal Gujarat and south Tamil Nadu experiencing wind speeds of 15-20 kmph. During the lean months of October to February winds exceeding 10 kmph are found only on the Gujarat, Konkan, south Tamil Nadu and Orissa Bengal coasts.

The finds of DNES shows that the Gujarat-Kutch coast, southern part of Tamil Nadu, coastal areas in Orissa, Anantpur in Andhra Pradesh and a few places in Madhya Pradesh have average wind speeds well above the limit for power generation.

Energy from the Sea

Basically, there are three ways of generating power from the high seas;

(1) Taming the waves (wave energy)

(2) Harness the tidal power (tide energy)

(3) Using the difference in temperature between the layers of the ocean.

Wave Energy

According to one estimate, the total global wave or tidal energy potential is around 3 million megawatts. Out of this India's coastline alone could generate a possible 40,000 megawatts. The reason for this? Traditionally, most ideal locations for utilizing wave energy have been the trade wind belts like the *Arabian Sea* and the *Bay of Bengal* Wave power culd well be one of the viable alternative sources of energy long after coal and oil run out.

Wave Energy in India: Recently, the country's first pilot plant that converts wave energy into electricity was inaugurated. Located off the Vizhinjam fisheries harbour near Tiruvananthapuram the device which has a peak generating capacity of 150 kw was designed by the Ocen Engineering group at IIT Madras.

The device operates on what is known an Oscillating Water Column (OWC) principle: A column of water rises and falls inside a concrete chamber, placed in the ocean in concrete with the waves outside. The action of the water column rising inside the chamber causes the air inside it to be expelled forcefully upwards and in the process rotates a turbine placed on the top of the Chamber. The turbine is coupled to an induction generator which produces the electricity. Similarly, in a reversal of process, as the water column falls, air is sucked into the chamber from outside which again rotates the turbine, thus maintaining continuous generation of Electricity. With the plant fully operational, the Department of Ocean Development (DOD) expects an average output of 75 kw during April-Nov. period, 25 kw during non-monsoon months of December to March and 120 kw during monsoon months, what makes the Rs. 2 crore Indian experiment interesting is the fact that almost all the crucial components of the plant were developed indigeneously.

High Cost: The economics of generating energy from waves is depressing at the moment the captial cost of a wave energy system is estimated at Rs. 1 lakh to Rs. 2 lakh per kilowatt of installed capacity compared to a mere Rs. 6500 to Rs. 13000 per kilowatt for a conventional one.

Tidal Energy

The term tide is used for the periodic rise and fall of waters of the ocean and produced by the attraction of the moon and the sun. The tides ocean about energy 12 hours. The tidal waves result by the gravitational pull on ocean water by the moon and the sun. Tidal flow consists of the vertical motion of the rise and fall of water level and a horizontal movement in the tidal currents towards and away from the shore. There are already tidal energy installations in the USSR (near Murmansk) and France (The Rance estuar) the latter providing an electricity output of upto 25 mw. Besides being available in practice at only a few sites, tidal energy has the economic disadvantage of requiring large capital expenditure and a long lead time before power is actually produced. It is therefore unlikely to make a substantial contribution to any country's energy supplies in the foreseable future.

Other Countries: The device or the technique employ to generate energy — is not particularly new. Research on similar lines has been going on for severe years now in countries like the U.S., the U.K., Norway, Japan, Sweden, Germany, France, Finland and Ireland. Japan has commissioned a plant called Kaimai Bnoy—Japanese effort which is the largest in the world has a rated capacity of 1 MW and apart from merely producing electricity from the waves also separates at uranium from the sea water while it is at it. Britain also inaugurated its first pilot plant, sometimes last year. Like its Indiana counterpart the British device is also based on OWC principle but unlike the Indian system the power modules one placed horizontally.

Norway put its first plant out to sea as early as 1986 only to have it blown away in a cyclone but another Norwegian company soon developed a system which is called Tapchan — from tapered channel. Different from most other designs the channel is a concrete slope that tapers as it rises three metres above sea level. The waves enter the channel and is pushed upward by the momentum of the waves behind it. The water finally collects in a reservoir at the end of the channel, which feeds a conventional turbine and the water is returned to the sea.

Ocean Thermal Energy Conversion(OTEC)

This operation exploits the temperature difference between the surface of the sea and at a depth of 1,000 metres or more to extract energy. In tropical countries such as India the strategy works even better as the temperature gradient in the seas is as great as 25 degrees centigrade.

Technique: Cold water is drawn up through a vertical pipe to condense a working fluid such as ammonia in a closed cycle. The warm water at the surface is used to evaporate the fluid. By pumping the working fluid around a closed circuit, the latent heat can be extracted through heat exchangers and used to generate electricity. This is called the closed system. A variation — the open system is also being tested. In this it is the warm sea water which is used as the working fluid. The warm water is passed into a vacuum chamber where it evaporates rapidly. After this, it can be condensed by surface contract with the cold deep sea water.

A third system — their reverse OTEC —which uses the temperature difference between the warm ocean and the cold air above it is also being pursued in the USSR. But the process is still under wraps. The oil crisis of the early seventies was largely responsible for catapulating the OTEC into lime light. However, the technology had been around much before that. Prototype OTEC devices had already been constructed in Cuba in 1929 and in the Ivory coast in 1956. The next attempt was only in 1979. This was a 50 KW plant sited oft Heavier and set up as a research tool for designing 1 MW units. Now, the U.S. is planning larger units with capacities of 40 MW.

Still there are several snags to be overcome before the bountry can be handled out from the sea. The main hurdles to overcome:

(1) designing suitable long cold water pipes

(2) advanced heat exchangers

(3) large pumps and

(4) overcoming biofouling in the system.

It is still too early to judge the economics of OTEC technology. But some constituency it will become competitive once oil prices touch it 25 a barrel.

Geothermal Energy

Vast quantities of heat are stored away in the lower layers of the Earth's crust. The temperature of the Earth's interior increases with depth and at most sites temperatures suitable for use on economic basic are focused only at conciderable depths. A temperature of 300° C (5720F) normally occurs only at depth of about 10 km (5 miles). In geologically recent volcanic zones and in fissures in the Earth's surface however temperatures of 300° C are not uncommon at lesser depths. Geothermal energy can either be used as steam to produce electric current in power stations or be used directly as primary heat.

Use of Geo Thermal energy in Various Countries: Scientists in Germany are working on a project for the harnessing of geothermal enrgy on the upper Rhine near Laudare. There at a depth of 1000 metres (3280 feet) the prevailing temperature stands at 90° C (194 F). The process is simple; cold water is pumped down to a given depth, where it is heated up by thermal flow and piped again to the surface. There it is used for domestic heating, to provide heat for industry and other purposes. California (U.S.A.) The Geyper power station in Northern California which are driven by geothermal steam power have an electric energy output of 180 megawatts. This they produce at a much lower price than comparable power stations driven by fossil or nuclear fuel.

Italy A geothermal power plant has been in operation at Lardevellow in Italy since 1904. Its output amounts at present to 370 Mvy. The steam which is emitted from boreholes under terrific pressure, comes from water that has seeped down from the earth's surface, probably mixed with sea water which flows in through newly formed fissures in the Earth's crust. The steam is piped into big turbogenerates and after serving its purpose is condensed in large cooling towers. From there it is run into chemical plants in which boric acid, ammonium and other chemicals are extracted from it.

Disadvantage of Geothermal Power Plant: Most of the geothermal power plants are relatively small having an output of between 10 to 20 Mkh. The heat reservers below ground have been too little explored so far and the exploration techniques are still at an early developmental stage. For these energy sources to be utilized on a large scale, various problems have still to be solved such as corrosion of turbines by water with a high salt content.

Nuclear Energy

Nuclear enrgy is quite a benign source — mainly because of the extreme care that is taken during design, construction and operation of the nuclear power plants. Nuclear fission has established itself as one of the potential solutions for this major problem. India has already gained considerable experience and expertise in all aspects of nuclear fuel cycle which includes.

(1) Uranium mining (2) Fabrication of fuel (3) Design (4) Construction and operation of nuclear power plant (5) Reprocessing of the spent fuel. An overall review of a such a programme including that from the environmental point of view is made.

Nuclear Energy in India: In the context of the enrgy science of our country nuclear power needs to complement the other sources of energy. India today has five nuclear reactors in operation. Two each at Tarapur (Maharashtra) and 2 each at Kota (Rajasthan and once at Kalpakkam (Tamil Nadu) with a combined installed capacity of about 1,095 mw representing about 2.5 per cent of the total power capacity in the country. Five more units are under construction at Kalpakkam, Narora (U.P.) and Kakrapara (Gujarat). The site of the sixth is yet to the announced.

The impact of surrounding air environment from nuclear power plant operations is different from those of conventional industrial projects. The atmospheric pollution from nuclear power plant during normal operations would be solely due to release of radioactive effluents. Several engineered safety features are provided to prevent and accident and mitigate the consequences should not accur.

As a matter of abundant caution off site emergency plans are also worked out for the unlikely event of an accident. Work on high level radioactive waste management relating to fuel reprocessing and also decommissioning of nuclear power reactors are continuing world over and current methods and processes for the same will evolve further and be improvised. All efforts are being made to improve the operating performance of our nuclear power stations.

INDEX